x + y

A MATHEMATICIAN'S MANIFESTO
FOR RETHINKING GENDER

EUGENIA CHENG

BASIC BOOKS
New York

Basic Books
Hachette Book Group
1290 Avenue of the Americas, New York, NY 10104
www.basicbooks.com

Printed in the United States of America

Originally published in June 2020 by Profile Books

First US Edition: August 2020

Published by Basic Books, an imprint of Perseus Books, LLC, a subsidiary of Hachette Book Group, Inc. The Basic Books name and logo is a trademark of the Hachette Book Group.

The Hachette Speakers Bureau provides a wide range of authors for speaking events. To find out more, go to www.hachettespeakersbureau.com or call (866) 376-6591.

The publisher is not responsible for websites (or their content) that are not owned by the publisher.

Print book interior design by Trish Wilkinson.

Library of Congress Cataloging-in-Publication Data
Names: Cheng, Eugenia, author.
Title: x + y : a mathematician's manifesto for rethinking gender / Eugenia Cheng.
Other titles: x plus y
Description: First edition. | New York : Basic Books, [2020] | Includes
    bibliographical references and index.
Identifiers: LCCN 2020014347 | ISBN 9781541646506 (hardcover) | ISBN
    9781541646513 (ebook)
Subjects: LCSH: Sex differences (Psychology)—Research—Methodology. | Sex
    role—Research—Methodology. | Social sciences—Mathematical models. |
    Categories (Mathematics) | Mathematics—Social aspects.
Classification: LCC BF692.2 .C454 2020 | DDC 155.3/3—dc23
LC record available at https://lccn.loc.gov/2020014347

ISBNs: 978-1-5416-4650-6 (hardcover); 978-1-5416-4651-3 (ebook)

LSC-C

10 9 8 7 6 5 4 3 2 1

To and for
the ever-changing congressive constant,
Gregory Peebles

CONTENTS

PREFACE

AS I WRITE this, the world is in a state of great uncertainty. It is the middle of March 2020, and a global pandemic is in the process of closing down society as we know it.

It seems impossible to begin this book without mentioning that, but also impossible to know what to say that will still be relevant when the book reaches your hands. The situation is developing faster than we can keep up with, and we don't know what is coming in the approaching days, weeks, months. The range of possibilities is unthinkably wide.

However, one thing that this worldwide crisis has grimly illuminated is a sharp opposition between those who think as individuals and those who think as a community. We see individuals deciding how much risk of infection they are going to take, as if this decision is their own private one, affecting them alone. What we need is individuals acting on behalf of the community, to reduce the risk to the community and protect our collective health.

This contrast between individualistic thinking and community-minded thinking is the major theme throughout this book. It gives me pain to see it being played out so vividly during our current crisis, but it also galvanizes me.

This crisis wasn't there when I started writing this book, and I don't know what state the world will be in while you are reading this book. Whatever the case, I hope the book helps us think more clearly about these contrasting approaches to life, and helps guide us toward a better future.

x + y

PART I
GENDERED THINKING

CHAPTER 1
INTRODUCTION

BEING A WOMAN means many things.

Many of those things really have nothing at all to do with being a woman—they are contrived, invented, imposed, conditioned, unnecessary, obstructive, damaging, and the effects are felt by everyone, not just women.

How can mathematical thinking help?

As I am a woman in the male-dominated field of mathematics, I am often asked about issues of gender: what it's like being so outnumbered, what I think of supposed gender differences in ability, what I think we should do about gender imbalances, how we can find more role models.

However, for a long time I wasn't interested in these questions. While I was making my way up through the academic hierarchy, what interested me was ways of thinking and ways of interacting.

When I finally did start thinking about being a woman, the aspect that struck me was: Why had I not felt any need to think about it before? And: How can we get to a place where nobody else needs to think about it either? I dream of a time when we can all think about character instead of gender, have role models based on character instead of gender, and think about the character types in different fields and walks of life instead of the gender balance.

This is rooted in my personal experience as a mathematician, but it extends beyond that to all of my experiences: in the workplace beyond mathematics, in general social interactions, and in the world as a whole, which is still dominated by men—not in sheer number, as in the mathematical world, but in concentration of power.

When I was young I didn't have any female mathematician role models. There were hardly any female mathematicians in the first place, but also I felt no particular affinity with those I did encounter, and no special desire to be like them. I did grow up with strong, high-achieving women around me at all levels, though: my mother, my piano teacher, the headmistress of my school, the UK's prime minister, the Queen.

I worked hard to be successful, but that "success" was one that was defined by society. It was about grades, prestigious universities, tenure. I tried to be successful according to existing structures and a blueprint handed down to me by previous generations of academics.

I was, in a sense, successful: I looked successful. I was, in another sense, not successful: I didn't *feel* successful. I realized that the values marking my apparent "success" as defined by others were not really my values. So I shifted to finding a way to achieve the things I wanted to achieve according to my values of helping others and contributing to society, rather than according to externally imposed markers of excellence.

In the process I learned things about being a woman—and things about being a *human*—that I had steadfastly ignored before. Things about how we humans are holding ourselves back, individually, interpersonally, structurally, systemically, in the way we think about gender issues.

And the question that always taxes me is: What can I, as a mathematician, contribute? What can I contribute, not just from my experience of life as a mathematician, but from mathematics itself?

WHAT IS MATHEMATICS?

Most writing about gender is from the point of view of sociology, anthropology, biology, psychology, or just outright feminist theory (or anti-feminism). Statistics are often involved, for better or for worse: statistics of gender ratios in different situations, statistics of supposed gender differences (or a lack thereof) in randomized tests, statistics of different levels of achievement in different cultures.

Where does pure mathematics come into these discussions?

Mathematics is not just about numbers and equations. I have written about this extensively before. Mathematics does *start* with numbers and equations, both historically and in most education systems. But it expands to encompass much more than that, including the study of shapes, patterns, structures, interactions, relationships.

At the heart of all that, pumping the lifeblood of mathematics, is the part of the subject that is a framework for making arguments. This is what holds it all up.

That framework consists of the dual disciplines of abstraction and logic. Abstraction is the process of seeing past surface details in a situation to find its core. Abstraction is a starting point for building logical arguments, as those must work at the level of the core rather than at the level of surface details.

Mathematics uses these dual disciplines to do many things beyond calculating answers and solving problems. It also illuminates deep structures built by ideas and often hidden in their complexity. It is this aspect of mathematics that I believe can make a contribution to addressing the thorny questions around gender, which are really a complex and nebulous set of ideas hiding many things.

HOW DOES MATHEMATICS DO THINGS?

Mathematician Sir Tim Gowers writes of the "two cultures of mathematics," which he characterizes as problem-solving and theory-building.

It seems to me that in the wider world, mathematics is seen as all about problem-solving, and that theory-building is mostly unheard of or otherwise neglected. Of course, the two are not cleanly separate, and in the end I think mathematics is most satisfying when the two come together: where building a theory also solves some problems. But what does it mean to "build a theory"?

Mathematical theories are descriptive, not prescriptive. They describe something that we see happening at the root of a situation, but they are not only there to predict how the situation will unfold. More broadly, they are there to help us think differently about the situation, to illuminate it, to help us understand how certain aspects of it are functioning. The theory is abstract—we have disregarded some of the surface details in order to see what is happening inside.

In mathematics a theory often starts with an idea or a possibility, and is built up as a collection of conclusions about that idea, or consequences of that idea, or properties of that idea. Sometimes it's really just a reframing of an existing idea, but a slight reframing can lead to a monumentally different theory. If you stand at the top of the building formerly called the Hancock Center in Chicago, you can look across the city and see a vast jungle of "civilization" sprawling out to the horizon. But if you just shift your angle a tiny bit so that you're looking straight down Michigan Avenue, the jungle becomes a grid system that will suddenly spring up before your eyes in near-perfect alignment. Sometimes in life all it takes is a slight shift in perspective, and this is often how new mathematical theories arise too.

I am going to propose a theory, or possibly just a reframing, to solve a problem.

WHAT IS THE PROBLEM?

The problem I am going to address in this book is the divisiveness of arguments around gender equality. Not all arguments are like this, but too many of them are. Sometimes the arguments push people in opposite directions because of the way we talk about gender, and sometimes arguments are futile because there is little clarity about what exactly is being discussed. Take the word "feminism," for example. One problem right from the start is that there are so many different definitions of feminism—it means completely different things to different people. As a result, there is a certain amount of talking at cross-purposes. Some people use the narrowest possible definition so that they can obviously justify denigrating it. For example:

> *Feminism means believing that women are better than men and that men are all bad.*

Other people use the broadest possible definition in order to persuade everyone to support it, or even to convince themselves that they are working toward its causes no matter what they're doing. For example:

> *Feminism means believing that women have as much right as men to choose how to live.*

With some people's definition of feminism being highly restrictive and some other people's encompassing almost everything, it is a divisive concept even before it has even really gotten past the definitions.

Aside from these arguments about definitions, there are various reasons some people are anxious not to associate with the word "feminism" at all. Some people stereotype feminists as angry, man-hating, anti-family women, and this is an off-putting image for potential feminists of any gender. This highlights another source of the divisiveness of feminism: it inherently seems to separate women from men even as it is trying to overcome that separation. It's hard to argue about men and women without acknowledging that they are different; if they weren't different, we wouldn't refer to them as "men" and "women." And then we get distracted and absorbed by arguments about the ways in which men and women are and aren't different. That argument is a distraction and a detraction.

The problem with this lack of clarity is that it draws us into a meta-argument—an argument about what we should be arguing about. There is an argument that the only experiences of oppression shared by all women are by definition those experienced by the most privileged of women: white women (more specifically, rich white straight cisgender women). There are many problems other than, and possibly worse than, the problems of those women, such as racism, homophobia, wealth inequality. We can thus also be distracted by a competition about which issue is the most important and pressing, instead of addressing them all.

Who benefits from these meta-arguments? I suggest it is the people who currently hold power and who want to keep it that way. While women fight each other about what feminism should mean, and about which intersectional branch of feminism should be given the loudest voice, and about who is the most oppressed, and while men oppressed on the grounds of race, sexual orientation, wealth, status, upbringing, education, gender expression, physical strength, or prowess in sports fight to be heard too, while all this disagreement rages between people who feel disadvantaged in any way, the people currently holding power can rub their hands in glee and consolidate that power.

The problem of definitions and meta-arguments is something that mathematics is very good at sorting out.

A MATHEMATICAL APPROACH

In mathematics we propose a theory by making a basic definition. We illuminate the definition by exploring some key examples. We justify the theory by demonstrating in what way it is ubiquitous and helpful. It might be helpful because it enables us to solve particular problems. It might be helpful because it enables us to think more clearly. I will argue that the theory I propose in this book does both.

In math, things need to be ubiquitous *enough* and helpful *enough*. They don't need to be helpful in all possible situations; even being helpful in one situation is enough, although the helpfulness does need to be weighed against any sense in which it might also be obstructive. This is different from evidence-based arguments. A mathematical theory isn't judged by the statistics of what it achieves in a large randomized sample. When working with evidence we typically test something on a large sample and see if it makes a statistically significant difference. Is it better than doing nothing, on average? In the case of a new pharmaceutical drug, is it better than placebo, on average? In the case of differences between men and women, are there differences between male behavior and female behavior, on average?

Mathematical deductions happen somewhat differently. First of all, they proceed according to logic, not evidence. Second, they are more likely to ask: *Under what circumstances* does this make a difference? Or even: Under what circumstances *might* this make a difference? This is very different from asking whether something makes a difference on average.

It is important to remember that averages—whether mean, mode, or median—do not apply to individuals. "The average person" is not a real person. Saying that "the average person" does something-or-other certainly doesn't mean that most people exhibit that behavior, and it does not tell us anything concrete about an individual in front of us. This includes the amount of sleep people need (some people really don't need much), the amount of calories needed to maintain a healthy weight (some people need to eat a lot less than the official daily recommended amounts), whether doing exercise makes you feel better (according to research averages it does, but it doesn't work on everyone), whether or not you can win the lottery (it is so unlikely as to be impossible, and yet people do win almost every week).

It is often pointed out that individual experiences do not generalize to large groups, but the reverse is also true: the average experiences of a large group do not apply to individuals. So how are we to think about people?

Mathematical deduction often operates by something more like case study. Something worked in this one situation; what made it work? And can we figure out how to replicate it or get it to work more widely? This is the sort of thought process I will be applying.

Importantly, this is a sense in which talking about personal experience is valid, as long as you're not claiming that your personal experience is necessarily universal, statistically significant, or typical. It's a case study, and so one can ask, as mathematicians do: What made this work, and how can we build on this idea?

I will use many case studies throughout this book, high-lighting what I find meaningful about some of the brilliant people who have achieved things in the world. Many of these will be women who achieved things in their own way, not by emulating the heroism, bravery, record-breaking, fearless-ness, or athleticism of men. Sometimes this means they are undervalued in society, and I'll talk about that too. I'll talk about the work of mathematician Emmy Noether, scien-tists Jocelyn Bell Burnell and Rosalind Franklin, the activist Susan Burton and her advocacy for a justice system that is more humane rather than punitive, the more famous and less famous achievements of Florence Nightingale, the business principles of Dame Stephanie Shirley and Mary Portas, the career trajectory of Michelle Obama, the experiences of Pro-fessor Deirdre McCloskey in her "crossing" (as she describes it) from male to female.

As is manifestly evident, I am not a biologist, psychol-ogist, philosopher, historian, sociologist, gender theorist, biographer, neuroscientist. I am a mathematician. So I am going to write this as a mathematician, theorize as a mathe-matician, and explore as a mathematician.

I am going to propose a reframing of the entire discussion around gender. This consists of focusing on relevant charac-ter traits instead of gender, and has at its center some new terminology to help us get started with this new dimension. I will show a vast array of ways in which this can help us move forward, whether we are women or not. Like many mathematical reframings, it is in a way only a very small step further than what many brilliant people have written about and understood before me. But somehow everything seems to stop just short of proposing a new dimension and new vocabulary. Much has been written about the problem without proposing a solution.

My research field, category theory, also began with an apparently small step involving a new dimension and new vocabulary. And yet this had an effect on the entire way of thinking of contemporary mathematics. I think something similar is possible for how we think about gender. What I am proposing is not a mathematics of gender, but a mathematical approach to gender. I am going to use mathematical thought processes to rethink our whole approach to gender.

THE PROCESS OF MATH

Math is not just about calculating answers. In fact, the theory-building part of math isn't really about calculating answers at all. On the right is a sketch of some of the phases involved when mathematicians build a theory. Often a first step is spotting patterns. At a basic level the patterns might be patterns in numbers, such as the fact that all multiples of 10 end in 0. They might be patterns in shapes; for example, the following grid of triangles has patterns of bigger triangles of various sizes, built up from the small ones.

explore
↓
observe patterns
↓
wonder why
↓
theorize
↓
test
↓
refine
↓
perform abstraction
↓
prove
↓
generalize

Humans have used patterns throughout the ages and across many (or all) cultures as a way of building up large ideas based on small ones. Nature also uses patterns to build complex structures such as petals on flowers or the spirals of a pineapple. At root, patterns are about connections between different things, or between different parts of the same thing. We can then think of patterns more abstractly to include phenomena that aren't necessarily visual, such as patterns in behavior, which are essentially similarities between one person's behaviors at different times, between different people's behaviors, or more broadly between the behaviors of different animals or different communities.

Math is about making increasingly abstract connections between things. The abstraction consists of forgetting particular details about a situation in order to see some deep similarity. This is how making analogies works, but math takes it a step further than just declaring that a similarity exists; it moves up one level in abstraction by examining what the similarity really consists of, and treating that as a new concept.

For example, the equation

$$1 + 2 = 2 + 1$$

is analogous to the equation

$$2 + 5 = 5 + 2.$$

But in math we don't just leave it at that—we say

$$a + b = b + a$$

for any numbers a and b. This abstraction makes our point simultaneously less ambiguous and also more open to generalization, in the sense of broadening to include more examples.

Similarly, if we think about a pattern of women not speaking up in meetings, and female students not asking questions in class, the similarity at an abstract level could be summed up as "women not speaking up in mixed-gender environments." At this point we have not measured how prevalent this pattern is, we haven't found what causes it, and we haven't found out how to change it. But in identifying the pattern we have made a start, by shaving off extraneous details and homing in on what is really important. This is a crucial step in building a good theory.

The abstraction then helps as we wonder why a pattern might be arising. Math is all about asking why, and delving deeper and deeper for more and more fundamental answers. A superficial answer to *why* women are likely to speak up less in mixed groups is "because they're women." We could test that theory, and what we are likely to find is that there are women who actually do speak up, and men who don't. We might find a statistical link between being a woman and not speaking up. If the link is only statistical, however, then causation isn't fully determined. In abstract math we seek to go a level deeper. Can we look deeper to find a causation? *Why* does being a woman cause this phenomenon statistically? We might then find that rather than it being about someone being a woman per se, it's about relationships between people, and about how different people relate to each other. The idea of studying relationships rather than intrinsic characteristics matches a major advance in modern mathematics: the development of my research field, category theory.

THE IDEA OF CATEGORY THEORY

Category theory is thought of as a "foundational" branch of math because it looks at how math itself works. Before category theory there was set theory, which has a different ideology and thus very different technicalities. Set theory is based on the idea that the fundamental starting point of math is *membership*—that is, whether or not a given thing is a member of a particular set. This is a bit like parts of society being all about what "clubs" you're a member of, whether that is literal clubs like the old-fashioned and deliberately exclusive gentlemen's clubs, or more abstract clubs, like political "tribes" of people who believe the same thing more because of their perceived membership in that tribe than anything else.

   Category theory takes a different starting point: *relationships*. It is built on the idea that we can understand a lot about something or someone by looking at their relationships with those around them.

Here are some pictures of how those two points of view differ. A set can be depicted as a collection of objects enclosed within a boundary, where inclusion and exclusion are determined by some intrinsic characteristics:

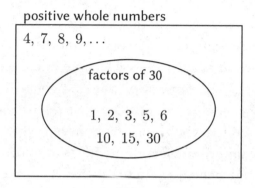

By contrast, a category can be depicted as a collection of objects with a network of arrows showing how they're related, a bit like in a family tree. In this case, instead of showing parent-child relationships we can show which numbers are related to each other as factors:

In a way this is only a slight change in perspective, but it took modern mathematics in a whole new direction. For our theory about gender, we can think of the set-theoretic approach as defining the set of women on earth according to some intrinsic characteristics. Sometimes this set is said to be determined by chromosomes, sometimes by reproductive organs, sometimes by hormones. In fact, none of these definitions is as clear-cut as some people assume, and furthermore the definitions do not match each other.

I will instead take an approach based in the ideas of category theory. This consists of thinking about how people relate to one another rather than what their biological descriptions are. Across history there have been attempts to define masculinity and femininity in terms of behavior, and how men and women relate to other people, and there are also definitions of gender as a social construct, as opposed to sex as a biological description. But in this book I will instead propose a theory that *only* looks at how people relate to one another, without trying to impose genders on those behaviors. These behaviors may or may not be statistically or biologically related to gender, but investigating that would be a different kind of study and is not really the point. Questions of whether certain behaviors are based in biology or society are a distraction; in a way, it doesn't matter, because all that matters is how people actually do relate to one another.

The idea of focusing on relationships *instead* of on intrinsic characteristics is an important part of the idea of thinking "categorically." I mean this in the mathematical sense of thinking according to the ideas of category theory, rather than in the non-technical sense of thinking in a fixed, clear-cut, immovable way. It's perhaps an unfortunate reuse of the word because, as I will explain, one of the important aspects of mathematical categorical thinking is flexibility.

The flexibility of category theory comes from thinking about relationships and freeing ourselves from thinking about intrinsic characteristics. This means that we have the potential to apply the ideas much more widely, to situations involving very different types of people, or even to zoom in and out and think about smaller-scale things like games and toys or about larger-scale things like schools, companies, and society more broadly. An example I wrote about in my previous book *The Art of Logic* involved thinking about power structures giving rise to structural privilege in society, such as being rich, or white, or male, or different combinations of them. The relationships we then consider are those between rich people and non-rich people, or between white people and non-white people, or between male people and non-male people. All those relationships combine into this diagram showing the relationships between people with different combinations of those particular types of privilege:

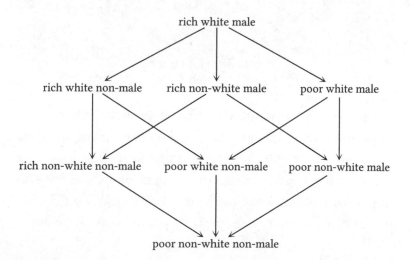

The arrows just depict hypothetical loss of one type of privilege, without making any claims about the causes or consequences.

As we are only thinking about the relationships between people who do and don't have those types of privilege, we can then apply the ideas to any other type of privilege among any other type of people, such as restricting our attention to women and thinking about the privilege of being rich, white, or cisgender.

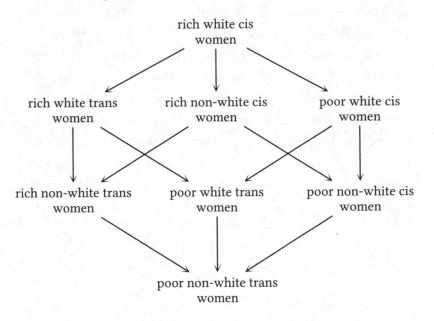

This abstract structure doesn't explain where those relationships come from, how exactly they manifest themselves in practice, or how to address the situation if, as I do, you believe it needs addressing. But it does enable us to gain some clarity about what the issues are, to focus on the issues that are relevant to a given situation, to package some relationships up into a single unit so that we can hold it in our brain more easily, and to move more smoothly between different situations looking for patterns. It is in this sense that I believe categorical thinking promotes flexibility of thinking, as well as depth of exploration and breadth of application. In fact, this is a key point of all abstract mathematics: a flexibility that comes from the act of ignoring some details of a situation, like shedding baggage so that you can travel more smoothly. In math it enables us to find unlikely similarities between very disparate situations, such as seeing that these diagrams of privilege are the same shape as the diagram of factors of 30 that I drew earlier.

But as with traveling light, you might worry about what you've left behind. There is a legitimate worry that we will lose something in the process of ignoring some details. Indeed, in arguments in normal life things often do get simplified to the point where important and highly relevant details are lost.

The key in math is that we choose which details to ignore for now, because of what we have decided to focus on, but that this is a temporary situation. We leave the baggage behind, but we don't burn it; we recognize that it could be useful for something else. It's a temporary abstraction in order to explore a particular aspect of a situation, not a be-all and end-all abstraction claiming to represent the entire situation.

With regard to gender, this is important because we are going to focus on aspects of the discussion that are only about how people relate to one another, not about intrinsic or biological characteristics; however, we are not claiming that the intrinsic characteristics are never relevant and can be forgotten forever. There are plenty of situations in which gender or sex really does play a role, whether it's because of biological aspects such as reproduction and other medical issues, or because of social aspects including prejudice and harassment, or because of statistical issues such as average body size and strength.

The idea of a temporary abstraction is to separate issues out and examine them individually, like in a controlled experiment where we try to vary only one thing at a time in order to see what is really causing what. If we conflate gender with character type, then we are not doing a controlled experiment, but rather we are mixing up issues.

The idea of a temporary abstraction also takes into account the thorny issue of intersectionality, and the need to take other forms of power imbalance into account besides those governed by gender, for example race, wealth, sexual orientation, and so on. In abstract math we don't do experiments, but we do study complex structures by trying to study one aspect of the structure at a time. This is in a way our version of a controlled experiment. The aim isn't to erase the other structures, but rather to see what is really caused by each of those structures.

As a basic example, if we try to study the world of numbers, we find that it is a very rich structure in which we can add, subtract, multiply, divide, and more. If we try to study all those things at once straightaway, we can become overwhelmed with trying to work out what is really going on. Instead we can study the idea of addition by itself, and see that subtraction is intimately linked with addition, so it can't really be separated out. Multiplication seems to be linked to addition, as it is often thought of as being "repeated addition"—multiplying 5 by 3 is the same as adding 5 to itself repeatedly: 5 + 5 + 5. But it can also be taken as a separate operation at a more abstract level, and this gives a wider possibility for application to multiplying things that aren't numbers, such as shapes, sets, or whole structures in category theory.

In this book my approach to gender will be to separate out issues in this way. The first step is to focus on gender issues temporarily, even though there are many other pressing issues. The next step is to notice that character traits might be thought of as being linked to gender but don't need to be, and so we can take them as a separate issue at a more abstract level. This gives us a wider possibility for application to things other than people, such as structures, systems, communities, arguments. In a sense, this is about introducing a new dimension.

DIMENSIONS

Dimensions, like patterns, can refer to more and less abstract things in math. At the more concrete level, dimensions are just different independent directions in which we can move in physical space. Our normal world is three-dimensional because there is the north–south direction, the east–west direction, and the up–down direction. (This is mildly complicated by the fact that we live on a sphere, but the principle remains, just with more care needed.) North and south don't count as different dimensions because they're not independent of one another (south is the "negative" of north), and similarly with east and west. Northeast doesn't count as a different dimension because it can be expressed in terms of north and east, so it's not "independent."

This idea of an independent new direction gives us a more abstract concept of direction in math. The "direction" can itself be abstract—so that instead of a compass direction for physical motion, it's an ideological direction for thought. If you measure how tall someone is and also what color their hair is, those things are independent in the sense that you can't state how tall they are in terms of what color their hair is. Category theory can be thought of as involving one more dimension than set theory because while objects by themselves are considered zero-dimensional, relationships between them make connections, like paths, which are one-dimensional. One example of this is what happens when we put numbers on a number line: they go from a collection of random zero-dimensional points into a one-dimensional line built from the relationships between the numbers:

$$\cdots \quad -4 \quad -3 \quad -2 \quad -1 \quad 0 \quad 1 \quad 2 \quad 3 \quad 4 \quad \cdots$$

We are going to introduce a new dimension to the discussions about gender, and that dimension is to do with the ways people relate to one another. I am going to think about ways of relating that I believe are particularly relevant to the inclusion of women in society, but we will approach this in a way that doesn't need to be directly tied to gender. Not only is it a different dimension from gender, I believe it is a different dimension from the existing ways in which we think and talk about character traits.

It is hard to define a new dimension because you can't express it in terms of the existing ones. However, this form of abstract invention is prevalent in math. It might seem a bit like making things up, but that doesn't have to be a bad thing. One example illustrating this is when mathematicians invented a new type of number called *imaginary*, by taking square roots of negative numbers. We sometimes say you can't take the square root of a negative number, but really we just mean that no ordinary number can be the square root of a negative number. So what about some other kind of non-ordinary number?

This is how mathematicians came up with the vividly named imaginary numbers. We basically just invent a number i that is going to be the square root of -1, so $i^2 = -1$. We don't know what i "is"; we just know its relationship with -1. We can also conclude that if the ordinary numbers go on a line, this funny new number i can't be anywhere on this line. So we might as well depict it in some new, independent direction, like this:

This is a sense in which we have invented a new dimension. You might think we just made it up, and moreover, we even acknowledge this by calling it imaginary. But we've made up everything in abstract math, really. We've made up the ordinary numbers 1, 2, 3, and so on—what in fact are those? They are just ideas. So we can make up the idea i too. We can make up any idea as long as it doesn't cause a logical contradiction.

The extraordinary thing is that this apparently fictional world turns out to shed a lot of light on our ordinary, familiar world. Sometimes it turns out that we have been looking at the world in too low a dimension to understand it, like if you look at shadows on a wall. Those shadows are a two-dimensional projection of a three-dimensional scenario, and it can sometimes be very hard to work out what the real three-dimensional situation is. In fact, sometimes the shadows can mislead us into thinking something else is going on entirely, such as in this picture:

which is actually a shadow of my hand in a contorted position:

Misleading "shadows" also occur with abstract dimensions, including the one I will introduce to rethink gender. We have been examining questions of gender inequality without this dimension, and I believe that this has been leading us astray; I believe that this new dimension, even though I sort of just made it up, sheds light on familiar situations. In fact, like category theory after set theory, this is in some ways only a slight change in perspective, but my belief is that it's a slight shift that can radically change how we see things, adding radical clarity and simplification of the best kind.

New dimensions are hard to work with because they're new, but it can be remarkably helpful in math to name them. The act of naming is another small step, but it can quickly help us organize our thoughts and then go further with them. New terminology is an idea that has been proposed by other authors on this subject, but I am going to go a step further and actually propose some words. You can have the thoughts without the words; you can also have the thoughts with different names. But it would have been hard to develop category theory without the word "category" for the new type of structure we're thinking about; it would be hard to study the made-up square root of –1 if we had to keep referring to it in that long-winded way instead of by its new name, i; and I think it is hard to proceed with our arguments on gender while we're stuck with our old and often gendered terminology. Naming a new theory is a starting point for making progress with it.

HOW WE MAKE PROGRESS

The type of progress we can make with this new theory is broadly twofold, and it can be evaluated in these two broad ways. When we evaluate a theory in math there is a practical side, that is, what the theory enables us to do, and a theoretical side, that is, the generality and flexibility of the theory, and its ability to balance finding a broad range of examples while also still separating out pertinent issues. The practical and theoretical are linked, of course, and the theory behind the theoretical evaluation (if that's not too convoluted) is probably that these features are signs of a theory that will eventually enable us to do many things, even things that we don't know about yet. Category theory is an example.

One of the important things category theory enables us to do is develop more nuanced ways to think about sameness. If we think about intrinsic characteristics, then we are often stuck saying things are the same if they have the same intrinsic characteristics. But if we move to talking about relationships, we can allow for things and people who relate to others in the same ways even if they have different intrinsic characteristics. In some cases intrinsic characteristics might be important, such as if someone is a body double for an actor in a film—but even then they only need to be similar in certain terms. For example, the male stunt performer Gary Connery doubled for the Queen in the opening ceremony of the 2012 London Olympics, when the Queen was shown apparently jumping out of a helicopter and arriving by parachute.

When discussing gender issues we can get stuck in situations where we attempt to say that men and women are "the same" while also trying to address the fact that there are far more men than women in many areas of life such as math, science, business, politics, wealth. Some people simply shrug and insist that men and women are different, so this imbalance is not unfair. Others agree that this situation is bad (because men and women are "the same") but still think we can't really do anything about it, because we must treat men and women "the same."

The categorically inspired approach will show us ways in which we can treat men and women the same if they relate to others in the same way. This doesn't mean that all men and women are the same, which is absurdly oversimplistic. It means that we can find the types of behavior that are important or beneficial, find people who exhibit those behaviors, and treat those people as "the same." For example, instead of saying something like "Men ask for pay raises more than women and that's why they're paid more," we could say "People who ask for pay raises are paid more" (regardless of whether they're men or women).* We can then go one step further and ask whether this is actually how we want the system to work: Should people be paid more because they ask for more? I believe they should be paid more according to evaluation of their actual work, not just because of their ability to ask for a higher salary. This provides a more nuanced solution, although it is also more difficult than simply paying women the same because "men and women are the same." But it is crucially less divisive.

The idea of sameness can be applied to individuals inside structures and also to structures themselves. This is a form of zooming in and out that is very typical of the sort of flexibility that category theory—and abstraction in general—gives us. Concrete things are too fixed to be used in this way. A book is a book, but if we go more abstract and think of a book as a collection of ideas communicated with some unifying coherence, then we can zoom out and apply it to libraries, and we can also zoom in and apply it to individual chapters or even individual paragraphs.

For the new dimension that we use to help us break out of the confines of gendered thinking, we will make it abstract enough that we can apply it not just to people but also to collections of people and to processes. We can apply it to activities, relationships, teaching styles, even different types of math. This is a benefit of this level of abstraction that I don't

---

*It is worth noting that there is evidence that women aren't paid as much even when they do ask for pay raises.

believe we can get from any of the existing words describing character traits that might somehow be related to our new dimension.

This is one of the theoretical ways in which a theory is deemed good in math, and it is one of the ways in which a theory has the potential to grow, as it can be applied so much more widely than is originally conceived. Take the idea of symmetry, for example. It starts as something we can see in objects that can be folded in half with both sides then lining up:

But if we think of symmetry more generally as a relation-ship between an object and itself in which we can some-how move the object and it still looks the same, then we include the idea of rotational symmetry, like windmills, or this shape, which is made of the same halves as the one ear-lier, but rotated rather than reflected:

If we expand this to other types of transformation on objects beyond just moving them, we get the idea of something that looks the same as itself if we zoom in or out, which gives us self-symmetry, such as in this fractal tree:

If we expand this further to include abstract transformations (as opposed to just physical ones), we get symmetry such as in this formula:

$$a^2 + ab + b^2$$

This has abstract symmetry because if we transform it by swapping a and b, it becomes

$$b^2 + ba + a^2$$

which is in fact the same as the original.

This is why a new theory in math can never be comprehensive at the start—or rather, if it is, then it must be a rather limited theory. A good theory has unlimited scope for expansion, just as category theory was originally applied to a very specific part of abstract algebra and then grew to include the rest of pure math, then theoretical physics and theoretical computer science, and more recently even biology, chemistry, engineering, and business.

Category theory has grown in this way by abstraction, by flexibility of thinking, and also by looking at things from different directions: from the outside in, and from the inside out. Sometimes we pick which relationships we want to study and see what world this makes, but sometimes we think about the kind of world we want to study and see what relationships will create this world for us. It's a bit like the internal and external approaches to math and indeed life at large: you can pick a path and see where it takes you, or you can pick a destination and then examine which paths will take you there. If the two agree, it's usually a sign that you're on to something good, and it can also be enormously satisfying. If the two disagree, then you need to decide which one you care about more. If you're trying to visit the Eiffel Tower but you see a very beautiful street that looks enticing, do you go down that street and risk missing the Eiffel Tower, or do you forge ahead and forgo the beautiful side street?

Abstract math is different from the concrete world because the things we're looking for can be dreams or dream worlds.

DREAM WORLDS

A fantasy world isn't exactly "real" (whatever that means) but can still help us shed light on the real world around us, just as fiction can (which is why it's so important to take interpretative art seriously). Dreaming about the sort of world we want to see can help us sort out what our fundamental beliefs are, as can zooming in and dreaming about the sort of life we want to live. For some people this means unlimited riches in order to own designer clothes, a private jet, and a yacht, but for others, including me, it means helping others understand things with the general aim of lifting up the disadvantaged. Those aims are rather different and entail rather different activities to achieve them. However, dreaming about what I would do if I did have unlimited riches helped me understand my fundamental beliefs and thus make changes in my life that I believe are beneficial both to me and to the world, even though I don't in fact have unlimited riches.

Math isn't just about getting the right answers; it's about dreaming up different worlds in which different things can be true. For example, children (and adults) often ask if infinity is a number or not. The answer is that it depends on what world of numbers you dream up. Infinity is not in our ordinary world of everyday numbers, and if we just try to throw it in by "inventing" it naively, we end up with all sorts of contradictions and paradoxes. But there are many ways to dream up different worlds of numbers that can include infinity without causing those problems.

In category theory we can dream up worlds in different ways. We can decide what type of relationship we want to focus on, and see what world this produces. But we can also decide what kind of world we want to see, and then look for the kinds of relationships that will make that happen.

Seeking social justice, in whatever form we believe in, can be helped by both of those approaches. We can benefit from dreaming up our imagined utopia in order to help us understand the overall features we really care about, and then we can try to work out what small steps will take us there from our current situation. But we can also zoom in and think about what sort of interactions we care about between people, and then patch those localized features together to see what global structure will result. I'll do both things in this book.

Sometimes problems arise because people hold fast to the local features and disregard the resulting global features, such as when they insist on only evaluating people by their achievements (not taking into account any life disadvantages such as gender, race, or socioeconomic background) and disregarding the fact that this perpetuates the underparticipation of disadvantaged groups in society. If local behavior that seems fair results in global outcomes that are not fair, I believe that, rather than just disregarding the global consequences, we need to make a conscious choice about how much that matters to us, like with the Eiffel Tower and the enticing back street. If dreaming about utopia and thinking about local issues take us to the same global outcome, then so much the better. I believe this to be true of the theory I am proposing.

I will end the book with my dreams for a world enlightened by these new ways of thinking about gender. I must acknowledge that for many people gender bias is not the most life-threatening problem they face. But I believe it is still a serious problem in the world, despite all the wonderful work of people who have gone before me. And I believe in trying to help the world in ways that we can, while recognizing our limits and always trying to do better at working toward the world we want to see.

I want to see a fairer world, in the sense of people not being held back by features about themselves that should not be relevant. I think that gender is often relevant in practice when it shouldn't be, and I dream of a utopia in which we focus on relevant character traits instead of gender and in which we do not associate those traits with genders, as we have so often done historically—for example, with traits like strength, ambition, confidence, empathy, kindness, communication skills. I don't want to see men and women pitted against each other in a battle for dominance. I don't want to see "masculine" and "feminine" character traits evaluated against each other in a zero-sum game in which only one side can win and in order for one side to win, the other has to lose. I don't want to see women urged to be more like traditional men in order to succeed when there are other ways to achieve the same success, and indeed other ways to make worthwhile contributions to society instead of those traditional forms of success.

I want to see more unified discussions and unifying solutions to these issues, where we avoid separating men from women and erasing non-binary people in the process. The world isn't a dichotomy between genders. The world consists of all of us together. This book is my dream of how we can use mathematical thinking to build a better, fairer world, and the beginnings of a plan for how we might be able to get there, starting small and building up gradually to encompass and reform more and more of our flawed social structures.

In two-dimensional math we draw two-dimensional graphs of y against x. But y doesn't have to be pitted against x; instead, x and y can be added together to make a new dimension and a new way of thinking.

CHAPTER 2
THE DIFFICULTIES
OF DIFFERENCE

ARE MEN AND women innately different in some way? And if so, is it justifiable to treat them differently?

We are going to begin by examining these related questions and, more important, the sorts of arguments that are used to justify answering yes. We are going to examine and refute these arguments mathematically. We're going to show the weakness of arguments that suggest gender imbalance is "just the way of things." But in the end, rather than just refuting these arguments, we need to reframe the entire debate so that we can stop thinking about gender differences where they aren't relevant and stop getting involved in arguments that mainly serve the people who currently hold power in society.

Why do we persist in thinking about gender differences? I think it's telling to think about who benefits, when we think about why this research is even being done. Why is anyone trying to prove that there are innate differences between men and women in intelligence, scientific ability, competitiveness, or any other traits that seem to confer high status in society? One general reason to cling to the idea of innate ability is to give ourselves an excuse for not being good at something. If I claim that I just have no natural aptitude for sports, that gives me an excuse for being very, very bad at sports. Conversely, when people declare that I am very talented at the piano, that negates the thousands of hours of practice I have put in. People can declare themselves to be a right-brained, "creative" person, and use that as an excuse for being disorganized. They can boast of being a left-brained, "logical" person, and use that as an excuse for being insensitive. (This is in spite of the fact that the left-/right-brain theory has been largely debunked.)

The more invidious reason to claim that people are born with certain traits is to avoid having to help people do any better. This is a way of not having to address our prejudices. If we can somehow argue that women are innately less intelligent than men, then we won't have to address issues of inequality in education, science, business, politics, and every echelon of power. If "innate" biological differences are found, they become fodder for people who seek a pseudo-rational basis to maintain structures that discriminate against women.

If the arguments are about biology, what can math do for us here? Math gives us a framework for making justifications and also for evaluating them, so it gives us a way of assessing the value of any particular opinion. This is why math can be relevant to all sorts of things that don't appear to be obviously "mathematical." Mathematics is too often seen as being all about numbers and equations, in which case anything that does not involve numbers or equations appears to be not "mathematical." But I think anything that involves some sort of justification can be examined mathematically.

A mathematical justification is called a proof. It is like a kind of journey. It has a starting point, a destination, and a way of getting from the starting point to the destination using logical deductions. And so we evaluate it by thinking about the starting point, and thinking about the logical deductions.

In addition, a journey is often not just about getting there but also about what you can do when you get there. Sometimes you can't do much except admire the view and come back—say, if you climbed a mountain. But other times you arrive in a city rich in culture where you can explore, learn, and challenge your worldview.

We are going to use this approach to evaluate some existing arguments about gender differences, and then make a theory of how these arguments are flawed. However, these existing arguments are not stated quite like mathematical proofs, and so the first thing to do is to find the (attempted) logical structure of the argument and express it a bit more like a mathematical proof by reducing it to its bare bones. This process of stripping away outer layers is an important step in the mathematical process. The outer layers often obscure what the real structure of the argument is, a bit like sleight of hand, and so stripping away those layers often exposes the flaws in the argument. This is one of the reasons that math uses very precise language and abstractions, to leave less possibility for that sort of misdirection. It's a bit like the fact that it would be hard to carry a concealed weapon on a nude beach.

CASE STUDY

Here is one much-discussed argument about the gender imbalance in science and math, involving the idea of assessing people according to "systemizing" and "empathizing":*

Men's brains tend to be stronger in systemizing than empathizing, and systemizing is important in mathematics, so it is to be expected that there are more men than women mathematicians.

This looks a bit like a simple string of implications:

1. Being a man implies being better at systemizing.
2. Being better at systemizing implies being better at math.
3. Therefore, being a man implies being better at math.

Now, if these were valid logical implications of the sort used in mathematical proofs, then the conclusion would be correct. This is because in pure logic if we know "X implies Y" and also "Y implies Z," then it is logically valid to conclude "X implies Z." This sort of inference is exactly how complex logical proofs are built up from small steps, where

---

*J. Billington et al., "Cognitive Style Predicts Entry into Physical Sciences and Humanities: Questionnaire and Performance Tests of Empathy and Systemizing," *Learning and Individual Differences* 17 (2007): 260–268.

the steps all neatly fit together to form a continuous staircase from the starting point to the conclusion.

However, in the situation I've described here, they're not really *logical* implications. They are something more complex and difficult. The first step is a statistical observation, not a logical implication. It has been observed that men, on average, tend to be better at systemizing than empathizing, according to some proposed definition of these things. The next step, the idea that systemizing is important in mathematics, is somewhere between an assumption and an observation. The idea that it is important in mathematics sounds logical, but that makes some assumptions about what "systemizing" really means and what skills are really important for research mathematicians (as opposed to people who are very good at mental arithmetic or math exams). There are some observational studies backing up this idea, but in that case the result goes back to being an observed statistical correlation.

The fact that these are statistical observations then raises the question of whether the effect is something innate about men or something cultural. A more honest chain of argument would go like this:

1. Men have been observed to be statistically more likely to be stronger at systemizing than empathizing, for some very specific definitions of these words.
2. A correlation has been found between this notion of systemizing and becoming a mathematician.
3. Therefore, we might expect more men than women to become mathematicians.

This is a rather weaker conclusion, reflecting how weak the steps in the argument actually are. It tells us nothing about whether or not it is fair or biologically inevitable that the gender imbalance persists.

I am now going to propose a general theory of these weak arguments, based on the preceding blueprint.

A THEORY OF WEAK ARGUMENTS

One important step in the mathematical process is to make a general theory that can then shed light on more than just one situation. We often do this with the help of abstraction, stripping away some external details to show the bare bones of a situation, which can then be seen to be the bare bones structure of other situations. This was the point of my introducing the letters X, Y, and Z in place of some parts of the argument used earlier—to focus on the logical structure of the argument that did not really depend on the details of what X, Y, and Z actually represented in this particular case.

Having done that to show what a sound logical argument would look like, we can contrast it with what the weak, unsound argument looks like, which is something like this:

1. Men are observed to have quality Y on average, under some select circumstances.
2. Quality Y is believed to be good for activity Z without any very strong basis.
3. "Therefore," men are naturally better (or worse) at Z.
4. "Therefore," we don't need to do anything about the imbalances in favor of men in activity Z.

We will be focusing on issues of gender in this book, but it's worth noting that this general argument form is applicable very widely to many situations other than gender where arguments about imbalance are raging, including disagreements about race, wealth, educational background, sexual orientation, and so on. One advantage of abstraction is that it helps us to see connections between a broad range of situations beyond the matter directly under consideration.

Anyway, the weak argument gets subtly but invalidly morphed into one that seems much stronger via a series of sneaky slides, as in the preceding example. "Men are statistically more likely to be better at systemizing than empathizing" turned into "Being a man implies being better at systemizing," involving some unsound deductions about statistics. The abstract version of this slide is something like this:

men have quality Y on average
↓
men have quality Y

There's another slide that turns "Men are observed to be better at systemizing" into "Men are by nature better at systemizing," assuming the effect to be a result of nature, not nurture. This is the sort of deceptive argument that enables some people to assert that gender differences are biological and therefore gender imbalances in the world are not the fault of discrimination. The abstract version is like this:

men are observed to have quality Y
↓
men naturally have quality Y

And then there's the slide that turns "Men are better at systemizing" into "Men are better at math," where the thing that has (supposedly) been measured is taken as a proxy for something much harder to measure. The abstract version is like this:

men have quality Y
↓
men are better at Z

where Y has been casually swapped for Z without much justification or fanfare. These three surreptitious slides can be combined to make arguments dramatically weaker by means of these less noticeable increments. This means that although we start at the top of the following diagram, we

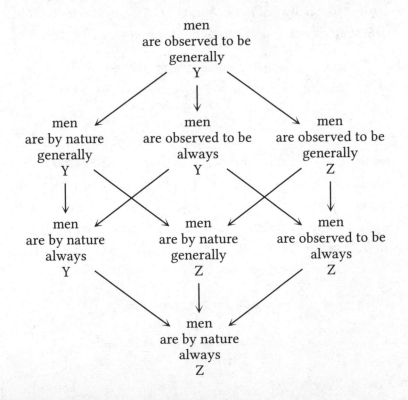

can sneakily claim that we are anywhere further down it by sliding down the arrows, but each time we move along an arrow the argument becomes more flawed.

The generality of this theory means that it can be applied to a wide range of examples where gender imbalance is found. In mathematics a theory is judged by the breadth of examples it unifies and the amount of light it sheds on those examples, so after making a mathematical theory we typically test it by trying it out on some more examples. We could try applying it to another type of argument that has been used to justify gender imbalances in academia, this time in physics:*

1. Men have more academic citations than women in physics.
2. Citations are a measure of how good you are at physics.
3. Therefore, men are better than women at physics.
4. Therefore, it is fair that there are more men than women in physics.

The first point is fairly well documented, but the second assertion involves less of a slide and more of an enormous leap of faith. The conclusion "Men are better than women at physics" may well be true statistically if we take a snapshot in time right now and take "better at physics" to mean more successful at making progress in advancing theories, but concluding that this is a fair situation is another giant

---

* See, for example, A. Strumia, "Gender Issues in Fundamental Physics: A Bibliometric Analysis," preprint, 2019.

unjustified leap: men might be more successful because the world favors them unfairly. In this case the issues are less about the observations being taken and more about the conclusions. We will discuss those issues, relating to the bottom right part of the diagram, in Chapter 3, and focus on the top left part of the diagram for the rest of this chapter: the part of the argument that is supposed to show us that there are some innate differences between men and women.

There is a strong perception of differences between men and women, and understandably so—there are some fairly obvious general differences between men and women physically. But we are going to discuss the flaws of taking those differences too seriously or concluding too much about those differences. Instead of asking whether or not gender differences are innate, it is more productive to ask in what sense they are innate, to what extent they are innate, and what the point is of basing our world on those differences.

When it comes to making arguments about gender differences, we have seen some flaws in the logic used earlier. There are also flaws of method, for example, in how we measure our observations in the first place, and flaws in how we understand the statistics even if some robust observations have been made. We will start at the beginning with the observations.

PROBLEMS WITH OUR OBSERVATIONS

Sometimes our conclusions based on observations go wrong because we just base them on our own experience. This could be unsound because we have a small sample that is skewed by our own situation. For example, I happen to know an unusually large number of women mathematicians.

It is even more unsound if we have succumbed to confirmation bias, where we tend only to notice evidence that supports our theory and not evidence that refutes it. This common type of bias is well known and well documented, but many people who seem to know about it intellectually still fall prey to it in practice. A male math professor saw me checking my academic hood in a mirror when we were all lining up for graduation, and said, "See, women are just vain."

Personal observations are prone to these sorts of errors, but science is supposed to be based on a much more rigorous and unbiased process of observation. Experiments are controlled, sample sizes are large, and the replicability of scientific experiments together with peer review is supposed to ensure that the observations are impartial and reliable. Unfortunately, this has not been the case when it comes to the search for differences between men and women. The prejudice-based flaws have been analyzed very comprehensively in books such as *Inferior* by Angela Saini and *Testosterone Rex* by Cordelia Fine. Note that this does not mean that science itself is flawed—sometimes people see any flaw in a scientific result as evidence that science is not "trustworthy" and we might as well go back to relying on personal opinion. The scientific process is a *process*. Part of that process is a process for finding flaws, and the fact that scientists are able to find flaws in scientific work is a sign that the process is working.

I will focus on the flaws that I can illuminate by means of logic and abstraction, rather than on the flaws in the experimental methodology. I will begin with the issue of trying to take observations to show "innate" properties of people in the first place. One rather serious problem is how to ensure that any differences you're observing really are innate and not learned.

Imagine you are trying to design an experiment to investigate the innate differences between men's and women's brains. How are you going to separate out nature from nurture? One way to do this is by studying identical twins who were raised in completely different environments. However, if we're trying to investigate, say, mathematical ability, then we would need to find a pair of identical twins raised in different environments where one became very good at math and the other didn't. Or where they both did or both didn't, despite the differences in upbringing. This is a little far-fetched, even aside from the slight problem that identical twins will probably be the same gender.

The other popular approach is to try to do the research on children when they're young, so young that socialization and environment have not yet had an effect. The trouble is that those things have an effect almost immediately, so the studies have to be done on babies who are just days old. (People often interact with babies differently if they're told they're a boy or a girl, right from the start.)

The next problem is trying to make a controlled environment for a baby. You'd probably have to get a robot rather than a human to interact with the baby. Plus, how is the experiment going to be double-blind with respect to the gender of the baby? This means that the people doing the tests and recording the data must not know the genders of the babies. You would have to take the newborn babies from their parents, leave them all without gender identifiers in a separate room, and have the researcher come and collect the babies without interacting with the people who had taken them from their parents. The researcher would then take the babies in turn to a room where they would be tested by a robot. Now try persuading any parent to let their newborn baby take part in this study.

The next problem, if you've managed to pry some newborn babies from their parents, is: What tests can you do on newborn babies? You can't very well ask them to solve math problems. Thus the experiment has to test something extremely basic, usually just timing how long different babies look at different pictures.

If any differences are measured, then the experiment will tell us exactly this: that some babies looked at some pictures for longer than other pictures. In one much-cited experiment the babies were looking at a human face or a mobile.* In another it was pictures of a car as opposed to pictures of a doll.**

What are we then to infer about men's and women's brains? There is a rather large leap from babies looking at pictures for a few seconds longer to the gender imbalance in math professors at research universities. And yet this is the sort of inference that happens:

---

*J. Connellan et al., "Sex Differences in Human Neonatal Social Perception," *Infant Behavior and Development* 23, no. 1 (2000): 113–118.
**V. Jadva, M. Hines, and S. Golombok, "Infants' Preferences for Toys, Colors, and Shapes: Sex Differences and Similarities," *Archives of Sexual Behavior* 39, no. 6 (2010): 1261–1273.

1. Girl babies looked at the face for slightly longer than the boy babies on average.
2. Therefore, girls' brains are hardwired for empathy, whereas boys' brains are hardwired for systemizing.
3. Math is about systemizing; therefore, this shows that men are hardwired to be better at mathematics.

This involves sliding through all the sneaky ways to subtly change an argument, with one part going through an extra step:

boy babies looked at the mobile for slightly longer
↓
male brains are hardwired for systemizing
↓
men are hardwired to be better at mathematics

The preceding problems are about what we infer from the behavior of babies. There is essentially a trade-off between the certainty that what we are studying really is innate and not the result of "nurture" and the certainty that what we're studying is actually linked with some meaningful aspect of adult behavior. This trade-off is summed up in this diagram:

age of subject

certainty that what we are measuring is:
– – nature, not nurture
••• indicative of adult behavior

Aside from this tricky question of separating nature from nurture, there is also the difficulty of actually finding something quantifiable to measure. If we are trying to study the intelligence of men and women, how are we going to measure intelligence? Tests that produce clear answers are necessarily very restrictive and thus only test something extremely restricted. So how this relates to whatever we actually mean by "intelligence" remains a big question. IQ tests only test your ability to take an IQ test (on that particular day). Historically, brains were weighed as a supposed way of studying the relative intelligence of men and women. This measure is certainly unambiguous and objective, but using it to indicate intelligence presupposes a link between brain weight and intelligence. In fact, historically, scientists assumed that men were more intelligent than women, found that men's brains weighed more on average, concluded that brain weight must determine intelligence, and then used the fact that men's brains weigh more to conclude that men are more intelligent than women. It's a breathtakingly circular argument.

More specifically, we might try to study differences in mathematical ability in order to decide whether or not it is "fair" that there are so many more male math professors than female ones. But how do we measure math ability? If we use tests of mental arithmetic, then all we're testing is the ability to do mental arithmetic, and the question remains how, if at all, that is linked to one's ability to do research math. (Plenty of non-mathematicians are better at mental arithmetic than I am, especially those who use it every day, which I do not.) Attempts to link mathematical ability to the "systemizing brain" use a questionnaire with questions such as "If you see a mountain, do you think about the geological processes that formed it?" and "If you have an electrical problem at home, can you fix it yourself?" The answer for me is no to both, but all that says about me is how I feel about mountains and electricity, not how good I am at math. Even more tenuously, the questionnaire asks about my understanding of people in social situations and seems to conclude that if I am able to understand people, then this is an indication that I am bad at math, which is an egregious myth to be perpetuating.

This is one of the flaws in the study that attempted to measure physics ability in men and women by measuring the number of citations of their papers. The number of citations is definitely measurable, but there is only a tenuous link between that and how good a physicist is, and it's affected by the implicit bias that may cause people to cite male authors more than they cite female ones.

In short, there is a trade-off between how measurable something is and how meaningful it is likely to be in terms of actual behavior. This is summed up in this diagram:

Aside from these essential logical problems with trying to study gender differences by observation, there are problems even if the observations we made were sound. Inferring things from observed differences in behavior is fraught with problems, many of which have to do with misunderstandings of how statistics work. In the rest of this chapter we'll explore the ways in which statistical results are simplified in order to summarize them briefly, losing crucial nuance in the process. Unfortunately, the modern world is more about click-baitery than nuance, and so a headline stating that "men are better than women" at something is going to be favored over one stating that the distribution for men is a little further up than the distribution for women but that there is a large overlap. One thing we readers can do is educate ourselves about this and other nuances so that we don't continue to fall for these melodramatic black-and-white headlines.

These issues will be familiar to readers fluent in different types of average, standard deviation, and shapes of distribution, but as these are a widely used tool of misinformation, I have included some demystification here.

AVERAGES

If we're trying to make a sweeping statement about men being different from women, it might seem that we can put ourselves on more secure footing by saying "on average." But declaring things are true "on average" can still leave many possibilities for what is the case about individuals. The relationship between the "average" features of men and women and the behaviors of individual men and women is complex, and knowledge about individuals can't be reconstructed from just knowing the average. Men may well be faster runners than women on average, but that doesn't mean that every man is faster than every woman.

Averages and percentages conjure up traumatic memories of math classes in many people, and unfortunately this makes them an easily used tool for misinformation and misrepresentation. Some unscrupulous people do that deliberately to manipulate public opinion, while others might be well-meaning but misguided. The result is that statistics can be used to make people seem more different than they are, but greater mathematical rigor can help us see through that.

When we're thinking about a large collection of data we usually need some way to sum it up more succinctly so that it's not just a huge list of numbers. We want to be able to understand its general shape and compare it with other sets of data. There is always a trade-off between succinctness and loss of information. The more succinct your summary, the easier it is to present and grasp, but also the more information you will have lost in the process of making the summary.

Averages are a way of summarizing a collection of numbers with just one number. We are doomed to lose huge amounts of information and nuance in that process. Sometimes that information is critically important, and hiding it is a way to manipulate information in favor of a particular point of view. Different types of average focus on different aspects of the information and hide the rest.

Here is an example—admittedly an extreme one—to illustrate the point. The figures listed here are some hypothetical salaries in a hypothetical company employing five men and five women. The type of average called the mean is found by adding all the salaries together and dividing by the number of people. In this case I have calculated the mean for men and the mean for women separately:

	men	women
	$200,000	$10,000
	$200,000	$10,000
	$200,000	$10,000
	$200,000	$10,000
	$200,000	$1 million
mean	**$200,000**	**$208,000**

This company has one very highly paid woman, which skews the mean salary for women so it is higher than the mean salary for men. The company might well boast, "We pay women more than men on average," or "The average woman at our company is paid more than the average man." They might further shorten this to the more eye-catching "Women are paid more than men" in a linguistic sleight of hand.

I hope that hearing these descriptions and looking at some figures gives you at least a feeling of queasiness that something here is not right. I have seen cases like this where the data are produced, the statistics are analyzed, the description is made, and the case is declared closed: gender pay equity has been reached, and in fact it's now unfair because women are paid more. The women feel that they've been treated poorly, as they largely know that they are underpaid, but they also feel helpless because the "hard data" do not seem to back them up.

When I feel this sort of queasiness, this is a signal to me that something is amiss in the logic somewhere. It is then the job of my brain to work out exactly, *precisely* what is causing that pain—almost like a doctor using symptoms to pin down a diagnosis. If I find the root cause precisely enough, then I can fix it, and the queasy feeling goes away.

In our hypothetical gender pay example, the escalation of phrasing went like this, in small steps:

the mean salary for women is higher than
the mean salary for men

↓

the average salary for women is higher than
the average salary for men

↓

women are paid more than men on average

↓

women are paid more than men

It is popularly quipped that there are "lies, damned lies, and statistics," but it really isn't the statistics that are lying here: it is the interpretation or presentation of them. Once we understand how flawed the mean is as a measure, then we need not be fooled by the rhetoric accompanying these statistics. In fact, it's not a bad idea to be highly suspicious anytime anyone mentions an "average" in any context. There: the reaction of horror felt by those traumatized by having to learn averages in school is in this way quite sensible.

In the preceding example the problem arose from distribution of salaries for women being highly asymmetrical, and the discrepancy can be revealed by considering medians instead of means. The median is found by listing the numbers in order and finding the middle entry. The median salary for men is $200,000, whereas the median for women is just $10,000—so much for equity.

While the supposed "equity" in this case has easily been debunked, it is not always so easy. Here is a more nuanced example in which you might feel in your gut that women are worse off, yet the mean and the median are both higher for women than for men.

	men	women
	$100,000	$10,000
	$100,000	$10,000
	$100,000	$105,000
	$450,000	$105,000
	$450,000	$1 million
mean	**$240,000**	**$246,000**
median	**$100,000**	**$105,000**

The problem here could be detected by also looking at the 75th percentile and the 25th percentile, but I hope you can see (or guess) by now that I could construct a data set in which women were paid more according to all these measures but they're still in some sense worse off. The more percentile points you consider, the harder it is to hide inequalities, but after a certain point you really might as well look at the shape of the whole distribution.

SHAPES OF DISTRIBUTIONS

The picture of a whole distribution can give us a much better overview of a situation. Here are some examples where in theory the mean could be the same in every case, but the way the data cluster or spread is very different:

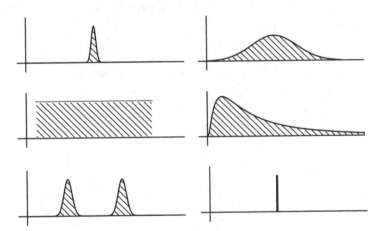

There is no type of average that tells us everyone behaves according to the average. That could only happen in extreme situations, as in the last picture, which just looks like a spike. There isn't even a type of average that means *most* people behave in that way. We have looked at the idea of mean and median, and it's important to remember that the mean doesn't represent anything very realistic in real terms except in select circumstances. If we're talking about the amounts that people are paid, as we did earlier, this would tell us how much everyone would be paid if we took the pot of money and divided it among everyone equally. But with something like height or "length of time spent staring at a picture," that's not very meaningful. However, the median does have a real meaning: that exactly half the results were bigger than this and half were smaller.

The mean will be in the middle if the distribution of data is symmetrical, like a normal distribution, shown on the left in the following figure. But on the right is an example where the mean is very far from the midpoint, which can happen if a few very extreme figures skew the results in one direction, as in our earlier example of the company with one very highly paid woman.

This last example is called the "log normal" distribution (the logarithm of the results has a normal distribution). This could be a graph of wealth distribution, for example. The superrich people skew the mean toward the right, but far more people have less wealth, which is why the peak is toward the left.

Many distributions of data are symmetrical, and in fact many display the sensible "bell curve"—the normal distribution, as depicted earlier, which peaks in the middle and tails off evenly to both sides. In that case not only is the mean the midpoint of the distribution, it also indicates a clustering so that most results really are clustered around that mean.

However, with different shapes of distribution the most common result could happen somewhere else entirely. The result that appears most often is called the mode. For the log normal distribution we can see how the skewing results in the mean, median, and mode being quite different:

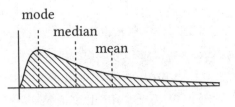

The mode is not very helpful where there is a very even distribution with roughly the same number of people achieving all results. In a situation with two peaks, none of those averages is at all useful.

This is called a bimodal distribution and indicates that there are two somewhat distinct subpopulations. This often seems to be the case in upper-level college math courses. Half of the students do very well and there's a peak in results at the high end, but half of the students struggle and there's another peak at the low end, with nothing much in the middle. It would also be what happens if we measured testosterone levels across the whole population: there would be a clustering and a peak lower down showing women and a clustering and a peak higher up showing men.

But often when comparing men and women the data are not completely separated out like that—there may well be two populations with different peaks, but with so much overlap that just talking about different averages is drastically misleading.

How much overlap there is depends on how "wide" the bell curve of the normal distribution is. These two pictures show pairs of bell curves where the peak is the same distance apart in each case, but there are very different amounts of overlap:

So just knowing the different means of the two groups is not enough for us to know how different the two groups are overall—that depends on how far apart the means are relative to how wide the bell curves are.

One way of measuring the width of the curve is the standard deviation, so named because it's a measure of how much the data points "deviate" from the mean. No matter how wide the bell curve is, there is always the same percentage of it within one standard deviation of the mean (on either side). That percentage is around 68. If the bell curve is more tall and thin, this is encapsulated by the standard deviation being smaller, and 68 percent of the results are then in a much tighter width around the mean. In the following picture the

dotted lines show the values that are one standard deviation away from the mean on each side; the distance between
the dotted lines is much smaller in the thin bell curve on
the right.

In the two pictures of overlapping bell curves shown earlier, the means differed by the same *absolute* amount, but by
very different amounts relative to the widths of the curve.
In *Inferior*, Angela Saini writes of a table "more than three
pages long" of all the statistical gaps that have been found
between men and women on all sorts of measures, including
mathematics, aggression, and self-esteem. She sums it up like
this: "In every case, except for throwing distance and vertical
jumping, females are less than one standard deviation apart
from males." She goes on to say: "On many measures, they
are less than a tenth of a standard deviation apart, which is
indistinguishable in everyday life." Here are pictures of two
normal distributions that are one standard deviation apart
(on the left) and one tenth of a standard deviation apart (on
the right), showing that in the former case there is already a
substantial overlap and in the latter case they are essentially
indistinguishable:

In *Testosterone Rex*, Cordelia Fine also goes into this, pointing out that there is not a "sharp line" between men and women but rather a "shifting mosaic of features." One meta-analysis of many studies found that "about 40 percent of the time, *at least*, if you chose a woman and a man at random, the woman's score would be more 'masculine' than the man's, or vice versa. (If there were *no* average sex differences, this would happen 50 percent of the time, so 40 percent is not far off that parity.)" Again, this included research on a wide range of measures such as mathematical ability, reading ability, competitiveness, and leadership style.

The mistaken idea that men and women are measurably different leads us to regard exceptions to this "rule" with anything from criticism to outright censure. When we try to sum up gender differences too succinctly, in effect we are making sweeping statements and casting some people in the role of "exceptions." The exceptions often turn out to be the result of our definitions or our assumptions rather than anything meaningful. If a female mathematician is considered an anomaly, does that tell us something about women, about mathematicians, or about our preconceived expectations? I would argue it's the last. It's true that there are many fewer female mathematicians than male ones, but perhaps it is this statistic that should be considered the anomaly. All too often this has to do with the assumptions we are taking as our baseline or default.

THE NULL HYPOTHESIS

One objection I often hear when I speak on this subject is the claim that we shouldn't be aiming for equality until we've proved *scientifically* that there are no differences at all. I think this is a question of null and alternative hypotheses.

In a scientific experiment, unlike in a mathematical proof, you have to start by deciding what is the default assumption in the absence of evidence. That is the null hypothesis. Then you look for evidence of something else going on. That is the alternative hypothesis.

For example, in a drug trial, the null hypothesis might be that the drug doesn't do anything beyond the placebo effect. The alternative hypothesis is that the drug does more than merely have a placebo effect. If you don't find evidence of anything other than the placebo effect, then you revert to the null hypothesis until you can find evidence to show otherwise.

In the case of gender equality, what should be the null hypothesis? When disagreements on gender bias and equality persist even among people who are not overtly prejudiced, they often seem to arise from a disagreement on that point. On the one hand, we could take the null hypothesis to be that there is no unfairness—that we should default to assuming that the numerical domination of men in certain fields is because of biology, not bias, until we find evidence otherwise. The anecdotal evidence of women's experiences of prejudice is not counted as evidence because it is anecdotal and not the product of a large randomized peer-reviewed study. Note that sometimes this is a way to hide overt prejudice in something that sounds justifiable and rational.

On the other hand, we could take the null hypothesis to be that there *is* unfairness—that we should default to assuming that the numerical domination of men in certain fields is because of bias, not biology, until we find evidence otherwise.

Who benefits from these arguments? Men benefit from the first one and women benefit from the second one.

Science is supposed to be impartial, but the choice of null hypothesis in the first place puts an immediate bias into any of these studies. Unfortunately, this is one way in which science has historically been used to hold women back.

And even if we did find that nature played a big part in gender differences in behavior, that causation is never going to be deterministic or measurable enough for us to really be able to make predictions based on it. That together with the fact that such a hypothesis benefits anti-egalitarian people convinces me that the argument is mainly a distraction and a detraction from the more pressing issues.

We have seen that even if men and women are different, there is a large range of behavior and large overlaps in the range of men's and women's behaviors. I think that instead of adopting a black-and-white position that men and women are different and thus gender imbalance is fair, it would be more reasonable to ask a more nuanced question about the range of gray in this area: To what extent are men and women different, and how much gender imbalance is thus fair?

However, we have seen the immense problems in the question of measurement. So even if men and women were innately different, the degree of difference might not account for the degree of gender imbalance we currently see. Moreover, it is likely to be more or less impossible ever to measure any exact degree of innate difference between men and women, so we will never be able to declare exactly what level of imbalance is "fair" as an outcome.

We could attempt to control experiments further and further in order to try to get closer to an "accurate" study, but in a sense this would actually take us further and further from real life.

LIFE IS NOT A CONTROLLED EXPERIMENT

We can bring even more nuance to our exploration by acknowledging that life is in fact not a controlled experiment. One group might be worse than another "on average" but have the potential to be better if nurtured and supported differently. One example is students who lack self-confidence. The received wisdom is that self-confidence helps you do better. However, students' lack of belief in themselves can lead to them being much better at recognizing their weaknesses and improving themselves, as well as being cautious enough to check their work thoroughly and back everything up with evidence and strong arguments; they just need more encouragement and support to get there. Whereas students who are very confident might be better at persevering in an unsupportive environment, but they are always in danger of putting forth shoddy work and baseless claims owing to their confidence. On average, all things being equally unsupportive, I would expect the self-confident students to do better. But in a *supportive* environment I would expect the self-doubting students to do better.

This effect is studied in *The Orchid and the Dandelion* by W. Thomas Boyce. The idea is that "dandelion" children are resilient and are largely unaffected by their environment, and that four in five children are like this. He calls the other children "orchids" and argues that, like the actual flowers, they are more sensitive to the environment and more likely to struggle in an unsupportive environment, but have the potential to do even better than the dandelions when nurtured appropriately.

Whether this is true or not, it exemplifies the idea that averages ignore the added nuance of asking in what circumstances different things happen. According to Boyce's theory, dandelions would do better than orchids "on average" if situations "on average" are not suitably nurturing (which they probably aren't at the moment). One response would be thus to favor dandelions; a different one would be to provide nurturing environments so that orchids too can reach their full potential to participate in society. Of course, dandelions benefit from not doing the latter, so self-interested dandelions are likely to oppose it.

This leads me to want to investigate whether men are more successful than women at certain things now just because of the environment that we have set up. This might be through things explicitly having to do with gender, such as prejudice, or through things that only implicitly have to do with gender via correlations and observed statistical differences, such as some character traits that are currently more prevalent in men and others that are currently more prevalent in women. I think this means that what is much more important than looking for gender differences is to look for character traits that are at the root of different people responding differently to different situations.

DIFFERENT ANSWERS IN
DIFFERENT SITUATIONS

Math is not just about getting "the right answers," because different answers can be right in different mathematical worlds. At a basic level, say in school math, someone else tells you which mathematical world you're required to work in, and then there might well be fixed "right answers." In the world of ordinary numbers 1 + 1 is always 2, you can't take square roots of negative numbers, and infinity isn't a number.

But in higher-level math it becomes more about dreaming up different worlds in which different things can be true, so instead of asking "Is this true or false?" we might ask "In what worlds could this be true and in what worlds is this false?" At this higher level of math you can find worlds in which 1 + 1 is 0, or 1, or 3. You can take the square roots of negative numbers if you move into the world of "complex numbers" instead of ordinary real numbers. Infinity is a number if you move into the world of the "extended real numbers."

Our current human world has been handed to us by past generations. In this world men and women might well appear to be different in various ways. Men are more successful than women, overall, in math and science and business and politics. But the mathematician in me says that there could be a different type of world in which that doesn't have to be true, and not just through us imposing quotas and insisting on fifty-fifty gender ratios. Not only are both of those divisive policies, but moreover they miss an important point: if something about the environment is preventing women from being as successful as men, then imposing a quota without thinking about the environment is going to increase women's representation without necessarily increasing their success. This is perhaps encapsulated as the difference between diversity (which is about numbers) and inclusion (which is about environment). But that still carries the danger of being stuck talking about gender differences, and thus being divisive and one-dimensional.

In Chapter 3 I'll look at the flaws in the sort of action we take in response to our ideas about gender differences, and then I will show that there is another approach we can take instead—by dreaming up a whole different world on a new dimension.

CHAPTER 3
THE PROBLEM WITH LEANING IN

HERE IS A kind of question that perplexes children in math lessons at school.

If Alex has 7 cookies and Sam has 3 cookies, how many cookies do we need to give Sam to make sure they have the same number of cookies?

Well, we could give 4 more cookies to Sam, or we could take 4 cookies from Alex, or we could make Alex give 2 cookies to Sam. In any of these situations Alex might be upset, depending on what sort of person they are. And in the math class the children might well wonder why we want to do this at all. Children are apt to ask questions that probe the validity of contrived math problems. What if Sam doesn't even like cookies and would rather have apples?

In higher-level math, if we're comparing x and y, then instead of asking how to add something to x to make it equal to y, we're more likely to investigate senses in which x and y are the same or different, and think about contexts in which things can seem the same or different. Three cookies and three apples are the same in terms of number, but not in terms of nutritional content. Depending on your tastes, they may not be the same in terms of pleasure either.

This requires us to ask deeper questions about what roles x and y are playing, and what aspects of x and y we're even interested in in the first place. These are much subtler questions that require flexibility of thinking, and often the ability to change perspective and find different abstractions focusing on different points of view. Higher-level math is not about fixed answers, but about such flexible thinking and deep questioning.

The issue of how to make things equal is relevant to gender differences. We have become so fixated on thinking about gender differences that we associate character traits with genders and then try to make genders equal by compensating for the perceived shortfalls on one side, like giving Sam more cookies. This is typically flawed twice over: first, in that we're associating character with gender in the first place, and second, in that we're assuming that the character traits associated with men are more valuable, and that to make men and women equal we thus need to get women to have more of the character traits associated with men.

I am going to argue that a better approach is first to decouple character from gender and then to think more deeply about the roles that different character traits play. Some aspects of our actual human experience are inextricably linked to gender when it comes to discrimination, explicit and implicit bias, structural power imbalances, and many instances of sexual harassment. But character types need not be associated with gender—in principle a man can perfectly well have some character trait that a woman has, and vice versa. I believe that if we decouple character from gender, we can think more clearly about the ways in which we currently overvalue certain character traits and about how we can change not just our assumptions but, crucially, also society's structures to value character types that we truly think are valuable.

This is hard work, and a big project.

GENDERED WORDS

The first part of the work is to stop automatically associating character with gender. At its most obvious, the assumed connection happens via the very words "masculine" and "feminine." What are these words doing for us? They are prescriptive rather than descriptive, in the following sense.

If you hear somebody (somebody more traditional, perhaps) describe a woman as "very feminine" or a man as "very masculine" or "manly," an image probably comes to mind, even if you wouldn't use those words yourself.* But why? The words do not just describe the behavior of women and men—they prescribe a supposed "ideal" or "natural" behavior of women and men.

Those terms might genuinely describe trends in men's and women's behavior to some extent, and at some point

---

*For an entire book about assumptions around these words, see *Femininity* by Susan Brownmiller (New York: Linden Press/Simon & Schuster, 1984).

in the past they might have done so even more. But across history in most (though not all) cultures, men and women have been expected to behave in certain ways, and we can hardly separate out those expectations from the behavior that then resulted. So to say that these gendered words really were descriptive of men and women's behavior in the past is to ignore this inextricable circular relationship between expectations and behavior:

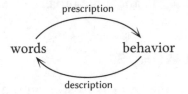

As we saw in Chapter 2, the idea of looking for gender differences is at best a distraction and at worst a tool of oppression. When it comes to "masculinity" and "femininity," both men and women suffer from pressures to behave a certain way. If a man is "unmasculine" or "effeminate," it sounds like something is not quite right, and similarly if a woman is "unladylike" or "masculine." We have fabricated a contrived system of judgments like this:

	men	women
"feminine behavior"	anomalous (bad)	appropriate (good)
"masculine behavior"	appropriate (good)	anomalous (bad)

We can free ourselves by thinking in terms of independent variables.

INDEPENDENT VARIABLES

In math, quantities may or may not have a definable relationship between them. When there is a fixed relationship it helps us reduce the number of things in the situation that we need to understand. This is often the point of an equation, such as this one for circles:

$$\text{circumference} = 2\pi \times \text{radius}$$

This is a definable relationship between the radius and the circumference, no matter what size the circle is, and it tells us that we don't need to understand those two concepts separately in order to understand a given circle, because we can deduce one from the other.

By contrast, if we think about a cone with a circular base

the radius of the base and the height of the cone are independent: we can't deduce anything about one from the other. There is no definable relationship between them. To give a less abstract example, in some jobs the salary scale is determined by how long you've been there, in which case you can work out how much someone is paid if you know how long they've been in the job. For example, if you have a fixed percentage pay raise each year, the graph of people's salaries might look like this, with the gray line showing that the situation is a one-dimensional line:

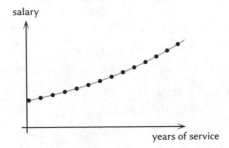

But in many jobs the salary depends more on negotiating pay raises, and so it can be wildly inaccurate to assume that people with the same experience are paid similarly (and, incidentally, neglects the fact that men on average apparently ask for, and are given, bigger pay raises than everyone else). The graph of people's salaries might look more like the one in the following figure—there is a vaguely upward trend, but any attempt to model it by a one-dimensional line (as shown) is going to be desperately approximate:

This tells us that instead of expressing salary *in terms of* years of service, we would do better to study those variables separately, instead of trying to use one to understand the other.

In more complex situations, an intellectual sleight of hand (sleight of mind?) happens in several small steps, making each one harder to detect, but compounding into a large misrepresentation (much as the small slides compounded into weak arguments in Chapter 2). One contentious example is body mass index, or BMI. The formula for BMI uses height and weight, so the relationship between height, weight, and BMI is defined and fixed. However, BMI is then taken as a proxy for body fat, which is then in turn taken as a proxy for healthiness, even though those relationships are not fixed or precisely definable. So we have the following chain of variables that we are considering:

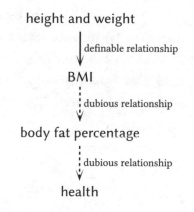

When we take one quantity as a proxy for another we are implicitly assuming that the two have a fixed relationship. When that isn't true, as in the case of BMI and health, or gender and character, we risk oversimplifying or even misinterpreting a situation. We can escape this by acknowledging that they are independent variables and considering the two separately. This means that we have moved into a higher dimension, because we now have to understand two things instead of one. For example, when we draw basic graphs in math we often study a relationship between x and y, such as

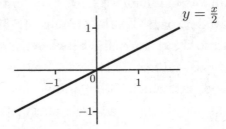

The graph of this relationship is a line, that is, it is one-dimensional, reduced from the two dimensions of the whole plane because of the fixed relationship between x and y. We have lost a dimension. Whereas if we acknowledge that x and y are independent, we get that higher dimension back— there is nothing to reduce our situation from two dimensions, and so we could be anywhere in the two-dimensional plane.

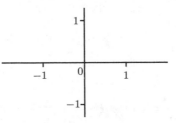

Going into higher dimensions makes things harder, which is one reason we might be reluctant to do it. But I think this is

what we need to do with gender and character: acknowledge that using one as a proxy for the other is flawed, and that we need to treat the two as independent variables. When gender is relevant, we need to consider it. When character is relevant, we need to consider it. But we shouldn't assume they are linked. In the rest of this chapter we will examine the consequences of *not* using the higher dimension, and in the second part of the book we will argue that the payoff for going into the new dimension is large and the effort involved comparatively small but that we are currently not caring to go there.

THE CURRENT ONE-DIMENSIONAL APPROACH

Currently, instead of considering gender and character separately, we are doing various confused things. On one hand, in the name of equality women are encouraged to be more like men in order to be "successful," and men are encouraged to get in touch with their "feminine" side, to learn empathy, express emotions, and be collaborative, although these are not gender-specific traits: people of any gender can have empathy and emotions, it's just that certain traits have traditionally been associated with certain genders, and we're beginning to see that that isn't necessary. On the other hand, there's a reaction to this that is sometimes referred to as "toxic masculinity," a response in which men, perhaps feeling disenfranchised by women encroaching on their roles, revert to an extreme version of traditional masculinity involving violence, often sexual violence in particular, and arguing for the reversion to male power roles and female submission. Here is a diagram depicting those tensions:

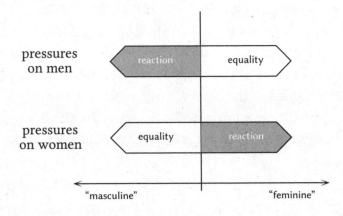

It is worth noting that if women become more like men, then certain men feel threatened. Unfortunately, I have long experience of this, since I am successful in mathematics, a field that is traditionally associated with men. I have seen men feel threatened by me in social situations, in professional situations, and in random interactions online. I once went to a party and tried introducing myself as working in a more traditionally female field (human resources), and had a much more friendlier response from men. It's true that this is just my personal experience, but if you pick any successful woman scientist on social media, I'm certain you'll see men feeling threatened by them.

In fact, this phenomenon has been studied in controlled experiments, with the results presented in articles such as "A Man's (Precarious) Place: Men's Experienced Threat and

Self-Assertive Reactions to Female Superiors."* Another study found that men felt attracted to women more intelligent than them as an idea, but were attracted to less intelligent women when it came to actually meeting them.** Unfortunately, I've experienced that rather a lot in the past too.

On the other hand, when men become more "like women" in the sense of being empathetic, communicative, in touch with emotions, then I feel supported, not threatened, and I'm sure it's not just me. However, I've seen other men feel it as a threat to the entire concept of masculinity, claiming it's a "war on men" or an "attack on masculinity." It is not a war on men; it's a war on the pressures of traditional gender roles, and in fact it's not even a war, but more of a gradual attempt to change culture.

A widespread response is that we need to redefine masculinity to give men a more clearly defined role in the new age of gender equality, or to persuade men that masculinity includes compassion and caring and respect for women. Actor Brad Pitt has been talking about his recent film *Ad Astra* in terms of rethinking the definition of masculinity, although

---

*Ekaterina Netchaeva, Maryam Kouchaki, and Leah D. Sheppard, "A Man's (Precarious) Place: Men's Experienced Threat and Self-Assertive Reactions to Female Superiors," *Personality and Social Psychology Bulletin* 41, no. 9 (2015): 1247–1259.

**Lora E. Park, Ariana F. Young, and Paul W. Eastwick, "(Psychological) Distance Makes the Heart Grow Fonder: Effects of Psychological Distance and Relative Intelligence on Men's Attraction to Women," *Personality and Social Psychology Bulletin* 41, no. 11 (2015): 1459–1473.

according to film critic Steve Rose the film is another in a long history of gender-stereotypical space films involving "heroic men and emotional women."*

There is a backlash from women as well. Some women say they like being "very feminine." But women can still be whatever it is that they currently think of as "feminine"; it just doesn't need to be associated specifically with being a woman. Unless it involves something like biologically giving birth to a child, men can do it too, whether it's wearing slinky dresses, styling your hair, making an art of makeup, or having an alluring way with men.** And if another woman doesn't do those things, then that shouldn't be called "unfeminine." Neither person should, in my view, have more or less claim to being female or woman-like; that idea keeps us stuck in one dimension, and worse, a skewed single dimension.

THE SKEWING OF THE SYSTEM

Being attached to a particular dimension or point of view isn't necessarily a bad thing, but in this case the gender dimension comes with embedded bias: inequality gets built into the system by a tendency to overvalue traits traditionally associated with men and undervalue ones traditionally associated with women. Take the debunked image of hunter-gatherers: we now know that around two-thirds of their food came from

---

* Steve Rose, "Fly Men to the Moon: Ad Astra and the Toxic Masculinity of Space Films," *The Guardian*, September 16, 2019.

** Some transgender men might be able to give birth to a child as well.

gathering and only one-third from hunting.* The hunt was dangerous and often unsuccessful, but—surprise—is still popularly celebrated and used to prop up justifications of male domination.

Fast-forward ten thousand years or so to the Second World War and Bletchley Park, the site of the famous code-breaking efforts that would eventually help defeat the Nazis. The work was spearheaded by Alan Turing and other mathematicians, but it required the collaboration of thousands of workers, around three-quarters of whom were women. Many of them performed routine clerical tasks, but others held degrees in mathematics, physics, or engineering and did specialist code-breaking work. Still, all of the work was crucial, including transcribing the coded messages, operating the code-breaking machinery, determining when a decoded message was recognizably German, and translating it. Of course, the leading mathematicians were mostly male because of how the education system excluded women; the rest of the workers were mostly female, as men who weren't expert enough for the leading math roles were away fighting.

It may well be that Alan Turing deserves the greatest credit because of the unique and unreplicable role he played in breaking the codes, but until quite recently a rather black-and-white situation persisted in which the men received all the credit and the women received none. Even when

---

*See Angela Saini, *Inferior: How Science Got Women Wrong—and the New Research That's Rewriting the Story* (Boston: Beacon Press, 2018), and Gerda Lerner, *The Creation of Patriarchy* (New York: Oxford University Press, 1986).

women's roles have been acknowledged, it has typically been as "support" workers to the men's roles of "genius." A lone eccentric male genius is of more interest than thousands of hardworking women. In *The Bletchley Girls*, Tessa Dunlop seeks to fill in some of the gray area by telling the women's stories from their point of view.

In the end Turing's achievements too were undervalued: he was gay and treated abominably by the British authorities instead of being hailed as a hero. Being male, it seems, is necessary but not sufficient to be considered a true "hero"; it is not only a patriarchy we are dealing with, and a white-ruled one, but a heteronormative one with many other oppressive qualities besides.

The leading mathematicians at Bletchley were selected through a drive to recruit "men of the professor type," especially from the universities of Oxford and Cambridge. It is not clear what the phrase "women of the professor type" would have meant to anyone at the time; Dorothy Garrod became the first woman Oxbridge professor only in 1939. As higher education expanded (finally) to allow women in, the image of a genius as necessarily male has perhaps weakened, but has it weakened enough? Recent data show that while many more women have careers in academia now than seventy years ago, the situation is still very unbalanced: there are far more men than women at more senior levels, and more women than men in part-time positions, which are automatically untenured and thus less respected in many places, including the United States.

But women may well be less respected regardless of their job title and achievements. Historically, high-achieving women are either less famous or remembered in selective, equivocal, or downright insulting ways. The 2013 *New York Times* obituary of the scientist Yvonne Brill began like this:

> She made a mean beef stroganoff, followed her husband from job to job and took eight years off from work to raise three children. "The world's best mom," her son Matthew said.*

Only in the second paragraph did the obituary note, "But [she] was also a brilliant rocket scientist." Not only did it lead with domestic descriptions, but it seemed to assert the improbability of being a good mother and a rocket scientist, inviting us to be surprised that anyone might be both. The *Times* quietly edited it after a general outcry, but only replaced the opening beef stroganoff reference with a reference to rocket science, and left the rest of the domestic introduction as it was. In fact, Yvonne Brill devised and patented a propulsion system for satellites that is still used today. The *New York Times* now runs a series of "Overlooked Obituaries" of people whose achievements were neglected in the past, typically because they were female or non-white.

A related example is Florence Nightingale, known as "The Lady of the Lamp" and arguably most widely remembered as a nurse. In fact, she was also a groundbreaking mathematician whose nursing was made effective by her analytical approach to data. Crucially, her innovative visualizations of

---

*Margaret Sullivan, "Gender Questions Arise in Obituary of Rocket Scientist and Her Beef Stroganoff," *New York Times*, April 1, 2013.

that data meant that her analysis was actually taken on board by those in power, enabling her to make the sanitary changes that would dramatically improve on the dire mortality rates in hospitals during the Crimean War. But "The Lady of the Pie Charts" doesn't sound as romantically evocative as "The Lady of the Lamp," and besides, she doesn't fit the popular image of a scientist as an old (white) man with mad hair—in fact, the exact image of Albert Einstein. Indeed, Einstein is much more famous than the mathematician who gave him crucial help in his work and who happened to be a woman, Emmy Noether.

EMMY NOETHER

Emmy Noether is a very famous mathematician among mathematicians and physicists but almost unheard of outside mathematics. (Perhaps that isn't surprising given how few mathematicians are heard of outside mathematics.) However, with the increased attention being paid to previously unsung heroes who were women, her prominence is rising, and she usually features on lists of brilliant women mathematicians, scientists, or "geniuses" we should have heard of.

Emmy Noether was at the forefront of women in mathematics in that she was in the first wave of women who were allowed to reach certain levels at each stage of her career. She was born in Germany in 1882 and attended Erlangen University at a time when women were only allowed to observe, not fully enroll as students. That had changed by the time she did her PhD, and she was then the first full-time female student at Erlangen. She went on to teach there—without pay, as women were not yet allowed to hold university positions.

Her work came to the attention of some eminent mathematicians in Göttingen: Felix Klein (of Klein bottle fame) and David Hilbert (who has Hilbert spaces named after him), who were working on Einstein's new theory of relativity and exchanging letters with him about it; those letters are very helpful for historians, and give us insight that we can't get when collaborations happen face-to-face.

They were stuck on a question about conservation of energy and they knew they didn't have the mathematical understanding to sort it out, so they called in Emmy Noether for help. Letters between Hilbert and Einstein indicate that her contribution was crucial and that they both respected and deferred to her expertise. At one point Hilbert includes a note from Noether to avoid a long explanation, and at another point Einstein writes, "How can this be clarified? Of course it would be sufficient if you asked Miss Noether to clarify this for me."*

Meanwhile, Noether was not allowed to have a teaching position because she was a woman, but for a while she gave lectures under the name of Professor Hilbert. The lectures were advertised as being given by Hilbert "with the assistance of" Noether, and everyone in the know understood that to mean that she was actually giving the lectures.

---

*See Yvette Kosmann-Schwarzbach, *The Noether Theorems: Invariance and Conservation Laws in the Twentieth Century*, trans. Bertram E. Schwarzbach (New York: Springer, 2011).

Einstein and Hilbert both petitioned for her to be given a teaching position, arguing that it was irrelevant that she was a woman because, as Hilbert famously pointed out, "this is not a bath house." There was resistance from other parts of the university, but eventually she was granted the lowest form of teaching position—not a real professorship, but still at least she was then paid (a little) instead of teaching for free.

This story, unfortunately, does not have a happy ending. As well as being a woman, Noether was Jewish, and soon the Nazis came to power and banned Jewish people from civil service positions. University posts were considered to be in the civil service, and so in 1933 Noether's position was withdrawn. She came as a refugee to the United States, where she was given a post at Bryn Mawr College, but there is still no happy ending: in 1935 she was operated on for a large ovarian cyst, and although the operation appeared to be successful, she developed a fever and died a few days later, at the age of just fifty-three.

So much for Noether's life and travails. What about her work? Her great contribution to physics and to Einstein's theory of relativity in particular was work connecting mathematics and physics—specifically, connecting symmetry in mathematics with conservation laws in physics. This insight did not directly solve the problem of conservation of energy that Einstein had, but it provided the illumination needed to help solve it. The question was, what happened to the laws of conservation of energy? Noether's work said: conservation laws in physics come from symmetries in the underlying mathematical system, so look for the mathematical symmetries and you will find the conservation laws. Noether's insight enabled a flow of understanding from one field of research to another.

Unfortunately for Noether, those sorts of moments are not as dramatic as the kind where you make a big leap forward, as Einstein's theory of relativity did. Noether was thanked and appreciated in many personal letters sent between Einstein, Klein, and Hilbert as they grappled with these issues. Klein wrote to Hilbert, "You know that Miss Noether advises me continually regarding my work, and that in fact it is only thanks to her that I have understood these questions." However, the extent to which they depended on Noether's help is less evident from the formal citations in the published papers. This might be technically correct if the theory of relativity did not directly depend on her work in a technical scientific way. Perhaps her work was more about shining light to help others find the way, rather than producing bricks and tools for them to use directly in their building. The people who produce bricks and tools are the ones who are officially credited, rather than those who shine light; is this fair?

Noether's work in pure mathematics also shone great light, and arguably was the beginning of a new sort of abstract mathematics that took off in the later part of the twentieth century. The birth of my field, category theory, was part of the journey toward greater and greater abstraction as a means of unifying mathematics. Category theory is sometimes criticized for building theories rather than proving big theorems, and sometimes it is criticized for "merely" shining more light on things that have already been proved.

Proving big theorems and solving problems that are hundreds of years old are a bit like the mathematical versions of climbing Mount Everest or going to the South Pole for the first time. There is something non-incremental about it, and thus we might call it "macho," falling yet again into the fallacy of attributing gender to a character trait. Once we attach the idea of maleness to a trait, we run the risk of overvaluing it and then asking everyone to emulate it.

EMULATING MALE BEHAVIOR

When patriarchal society overvalues character traits associated with men, there are broadly two common responses. One is to claim that gender differences are innate and that women thus can't expect to be the same as men; the other is to urge women to emulate men in order to be successful.

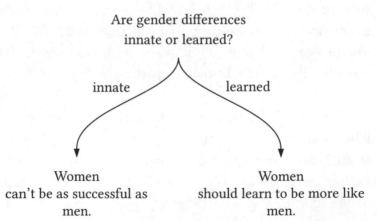

Men may claim that differences are innate in order to exclude women from male-dominated activities, and to relieve themselves of the responsibility of having to do anything about gender imbalances. But women may say it too: successful women may say it in order to retain their status as a rarity, and others to relieve themselves of the pressure of trying to emulate men or strive for success according to the existing definitions of success. After all, women may not want to emulate traditionally masculine behavior (and in fact men may not want to either). If the choice seems to be between emulating that behavior and not being successful, many people may choose the latter. But that choice is a false dichotomy, based on flawed underlying beliefs.

I have my own personal story of emulating the behavior around me to become successful at something predominantly done by men, that is, mathematics. My story goes like this.

When I started as a PhD student I decided I should hide all aspects of femininity as much as possible, so as not to give anyone a chance to stereotype me and say I was no good because I was a woman. Without being explicitly conscious of it, I started emulating behavior of successful people around me in order to be successful, and as the environment was male-dominated, that meant I was learning to emulate the behavior of men. But this doesn't quite make sense, as of course there is a whole range of behavior of men and a whole range of behavior of women.

In fact, after several years of university life mostly surrounded by male professors, I moved to a women's college as a junior research fellow and met a large number of women professors for the first time. I found that the environment was, disappointingly, not very different, because so many of the women took on the dominant and domineering roles that men might otherwise have occupied. The women were very successfully emulating stereotypically male behavior, and eventually I found I was doing it too: I was learning to be competitive, ambitious, assertive, unyielding. When I found myself in a high-powered academic environment, I discovered that certain people—not all—were prone to be argumentative, skeptical, dismissive, and instantly ready to jump up and challenge anything that anyone said. I tried to learn to do the same.

In *No Contest: The Case Against Competition*, Alfie Kohn describes "pseudofeminism" as seeking "the liberation of women through the imitation of men." It is a compounding of two assumptions: first, that gender differences are tied to character differences, and second, that the character types associated with men are more valuable, or even critical, for success. Of course, it's not just women who are imitating men; it's men too.

Emmy Noether, Florence Nightingale, Yvonne Brill, and the women of Bletchley are just a few examples of how women have been successful without necessarily emulating the behavior of men. Unfortunately, to counter the effect of women being valued only for traditionally acceptably "feminine" pursuits, there is a tendency then to celebrate them for being the opposite. There is a recent glut of books celebrating women who were "just like men," emphasizing that women can do everything men can do, or emphasizing traits that are not the traditionally "feminine" ones. Their titles include words like "headstrong," "badass," or "rebel," and I fall over them anytime I look for books about female role models from history.

Sheryl Sandberg's book *Lean In: Women, Work, and the Will to Lead* seems to urge women to adapt themselves to the (male-dominated) business world in order to be successful in it. It has been criticized in many ways, and one is that it seems to assume that in the current structures women can succeed if they just try harder, regardless of built-in sexism, racism, and other inequalities of access. And "trying harder" all too often seems to consist of trying harder to behave like successful men. I don't think our aim should be to show that women can be "just like men." I don't think we should have to imitate or emulate typical male behavior in order to be successful. Jessa Crispin includes this in her bracing take-down of some common forms of feminism: "a fight to allow women to participate equally in the oppression of the powerless and the poor."*

Making sure women can do all the things men do is some sort of progress but it doesn't necessarily represent or cause a great leap forward for all other women. Sheryl Sandberg's approach has been criticized as a "trickle-down" version of feminism.** Having Margaret Thatcher as prime minister didn't exactly bring about the end of discrimination against women in the United Kingdom, and the glass ceiling is alive and well despite her apparent shattering of it.

---

*Jessa Crispin, *Why I Am Not a Feminist: A Feminist Manifesto* (Brooklyn: Melville House, 2017).

** See, for example, Melissa Gira Grant, "'Like' Feminism," *Jacobin*, March 4, 2013.

Asking women to behave like traditionally successful men assumes that that sort of behavior is crucial for success. It is what I referred to in Chapter 2 as the assumption "Y is important for Z" where Y is some quality that men are perceived as having more than women do. Thus if women want to be successful, they must learn quality Y—that is, emulate the men.

WHAT TRAITS ARE VALUABLE FOR SUCCESS?

When I reconsider the behaviors I learned during my time in mainstream academia, I wonder if some of them were valuable. Perhaps being able to be uncowed by aggressively dominant people is a genuinely useful skill to have learned, but it's only useful because aggressively dominant people exist. If they didn't exist, it wouldn't be a useful skill at all.

In any case, I eventually hit a limit as to how much of this behavior I wanted to learn. Once I was tenured I felt more secure and started allowing some "femininity" in, but only outside work. This created a strange double life, so I tried to merge the two and be feminine at work too. That's the narrative I gave myself, anyway—I find it interesting to note that this meant something to me at the time, but I've now moved so far from associating character traits with gender that it's hard for me to reconstruct it. I think at the time "femininity" meant things to me such as showing vulnerability, not demanding to take charge of situations all the time, listening to other people more, and occasionally wearing a dress.

I realized that I didn't like the stereotypically masculine behavior, and didn't like myself when I behaved like that. I had been learning behavior I didn't like. But I couldn't work out how to unlearn that behavior and stay in academia; it felt as if the two were inextricably connected. Then I realized I wanted to take my career in a rather different direction, and over the course of a few years I changed my life around completely. I've built a new career in which I really don't have to encounter those types of people anymore, so I rarely have to invoke those skills. Even if I do encounter such people, I try to behave differently now, in order to defuse rather than escalate that type of interaction, and I wonder whether that competitive behavior I learned was important even for academia, where I learned it. Was there another way to be successful in that environment? More radically, was there another way for that environment to be?

When I was still in the process of changing my career around I had another experience that reminded me of my time in the women's college and galvanized me further. I met a woman at a party who mentioned her $5 billion budget in her first sentence, along with her name. She proceeded to be very combatively forthright about her success in a male-dominated business world, even going as far as boasting that she could compete with any guy when it came to "locker room talk." Her body language was not just strong but aggressive, and she seemed to invite people to challenge her just so that she could take them down a few notches, possibly by comparing the size of their respective budgets.

One might be tempted to call such a woman "masculine," and she might take it as a compliment. She did, after all, seem to have emulated "male" behavior in an expert fashion in order to rise in a competitive business world. I tried to point out to her that I have not had to emulate men's "locker room" behavior in order to be successful (we should also acknowledge that most men do not engage in so-called locker room behavior either). She replied with a sneer, "Oh yeah? And what makes you think you're successful?"

Although I was very bothered by this encounter at the time, I'm now grateful to the Five Billion Dollar Woman for prompting to me to think really hard about what I do consider to be "success." I started out assuming that I wanted to be successful via an already accepted route: getting a PhD, having my original research published in international journals, winning tenure, getting promoted. Eventually I realized that those definitions of success are a bit meaningless to me. I was expected to apply for grants in order to get promoted. But I realized I didn't really desire those promotions, and I didn't really need grants either. For me, success is about the people whose lives I have helped. As I am an educator, my biggest opportunity for helping people is in helping them to understand something they find difficult—for example, mathematics. This is a different type of success. It is a type of success that is not celebrated as much by society, and it took some effort for me to see that my values aren't the same as the ones the standard academic career path pushes.

SUCCESS IN MATHEMATICS

One traditional marker of success in mathematics is the Fields Medal, the prestigious prize for mathematics that is sometimes described as the "Nobel Prize of mathematics" (because there is no actual Nobel Prize in mathematics). No woman won the prize until 2014. The first woman to win it was Maryam Mirzakhani, and she was also the first Iranian to win it; tragically, she died in 2017 of breast cancer. She described herself as a "slow" mathematician, and she worked on problems by doodling on large pieces of paper, something her young daughter described as "painting."* This is a far cry from the idea of mathematicians as competitive machines, solving problems at speed.

A marker of "success" at an earlier stage in mathematics is the International Mathematical Olympiad, an international math competition for high school students. The training that some of these youngsters undergo for this competition is arguably as intense as training for the actual Olympics. And it is not that rare for the winning team to be all boys. According to the website FiveThirtyEight, in the history of US participation in the event (beginning in 1974), 88 percent of its six-person teams have been entirely male.** The United States won in 2018, 2016, and 2015 with all-male teams each time. (In 2017 South Korea won and had one female team member.)

There is ongoing debate about the lack of girls at high levels in this competition. Are they not chosen for the team because of bias, or are they just not as good at math? Some

---

*Andrew Myers and Bjorn Carey, "Maryam Mirzakhani, Stanford Mathematician and Fields Medal Winner, Dies," *Stanford News*, July 15, 2017.
**Leah Libresco, "Girls Are Rare at the International Math Olympiad," FiveThirtyEight, July 22, 2015.

people point out that they can't be chosen for the national teams because there are hardly any girls competing on regional or local teams. Why is that? How can we persuade more girls to take part?

But I would rather pose a different question: Is it important for us to persuade more girls to take part? The Olympiad is not the be-all and end-all of mathematical achievement. It is a constructed competition in a field that really doesn't have competition as its focal point. The Olympiad teams do often involve people who go on to become great mathematical researchers, but there are plenty of other ways to become a brilliant mathematician. In fact, math competitions run the risk of actually putting off some people who like math but don't like competitions—such as me. (Incidentally, Mirzakhani did achieve the highest level at the Olympiad in 1995, achieving a perfect score and two gold medals despite later calling herself a "slow" mathematician.)

I don't think we necessarily need to try to find ways to get girls to succeed at those competitions; I don't think that's the right aim. I would rather we identify worthwhile strengths that do not shine in competitions or are not recognized by the other traditional measures of success, and then find ways to encourage and value those strengths, regardless of anyone's gender.

Instead of getting women to imitate or emulate men, we can value them in their own right. Much has been written on the strengths that women bring to teams and to management, although—as usual—we risk making sweeping statements and having to qualify them by pointing out that not all women have those strengths and many men have them too. These are strengths such as communication skills, empathy, collaboration, which are valuable to any group of people trying to achieve something together.

However, valuing women for these "feminine strengths" and valuing men when they exhibit those traits does not address our fallacious assumption that character traits are tied to gender. It doesn't free us from the divisive gender debate. It doesn't take us out of one dimension. If we object to the idea that "men are better," it's not that helpful to declare instead that "women are better." It pits men and women against each other and sets up a prescriptive framework rather than a descriptive one.

In math we only use descriptive theories. There is no way for an abstract theory to impose behavior on concrete life, and so if our theory doesn't accurately describe something, then it might be a logically sound theory but it won't be relevant or helpful. Theories about people can influence people's behavior, especially if the theories are believed by a significant proportion of society, but that doesn't make them helpful; in fact, it makes them more destructive, where the mathematical version would just be irrelevant without really doing any harm.

Instead, we can do what math does when things have become a bit confused: we go back to first principles.

GOING BACK TO FIRST PRINCIPLES

In math, going back to first principles usually means working out what assumptions we are relying on, and shedding as many as we can. It means digging deeper through our thought processes to go back to much more fundamental concepts.

For example, when a woman is sexually assaulted, attention too often turns to what she was wearing. But if someone claims that she was sexually assaulted because she was wearing revealing clothing, they are implicitly assuming that men (as the assailant is usually a man) are unable to prevent themselves from assaulting a woman in revealing clothing. The following diagram shows the combination of those factors, with the dotted line showing the part that is too often overlooked:

To pick a less violent example, when I turn up for a speaking event wearing a dress, the tech guy quite often says that it will be hard to mic me because I'm wearing a dress; I typically interject "And because the equipment is designed for men" to draw attention to the baseline assumption that this is the only possible way microphones could be designed.*

---

*Note that the tech person is usually a guy but not always, and the non-male tech people I've had have never said this to me. Also, I now carry around my own gear that makes it possible for me to wear lavalier microphones designed for men even when I'm wearing a dress with no belt, no collar, and no pockets.

I am wearing
a dress.

Microphone equipment
is designed for men.

It is difficult to attach a
clip-on mic and transmitter to me.

Note that there is another assumption here, that men will wear clothes with a collar and pockets; we could have added in a dotted line connection for this.

Now let us examine the assumptions contributing to the idea that "character traits traditionally associated with men are better for success":

Society works in
a certain way.

Sucess looks
a certain way.

This type of sucess
must be achieved
a certain way.

There are gender
differences
in character traits.

Traits traditionally associated with men
are better for success.

In this case this means shedding the assumption that gender is tied to character and separating out those variables. This means focusing on gender when that matters, and focusing on character traits otherwise.

It's important to remember that in many senses gender does affect people's experiences, and experiences affect behavior. I am not saying we should be "gender blind." While inequities continue to exist in the system, and while historical inequality still has an impact on the present, I advocate taking much more active steps to counteract this. But the steps we need to take can, I think, be understood more clearly if we consider gender and character separately. The story of Dame Stephanie Shirley is a case in point.

DAME STEPHANIE SHIRLEY

Dame Stephanie Shirley was at one point the only woman on the *Sunday Times* Rich List—a list of the one thousand wealthiest people in the United Kingdom—so even the Five Billion Dollar Woman might deign to call her successful. She was a pioneer twice over, first in computer programming (which is how she made her money) and then in the care of autistic children (which is how she spent it).

I learned her story from her memoir, *Let It Go*. She and her sister arrived in London as refugees in 1939 when she was five, rescued from Germany on the Kindertransport along with thousands of other Jewish children. They were cared for by foster parents in the Midlands. She had a flair for mathematics and encountered gender-based obstacles early: her girls' school did not teach mathematics, and so she had to go to the local boys' school for math lessons.

She became a computer programmer in the 1950s, one of a very small number of them who happened to be a woman. She found it extremely difficult to be taken seriously or indeed hired, especially after having children, and she also writes of experiencing sexual harassment at work. So she decided to set up and run her own company from home instead, and specifically tap into the pool of neglected talent that women computer programmers represented.

At first she had some of the same problems—she couldn't get potential clients to take her seriously. So then she tried an experiment, sending out the same letters seeking business but now signing them "Steve" instead of "Stephanie." Work started coming in.

While this might sound like emulating male behavior, it's really just using a male name in order to overcome bias, not unlike Charlotte Brontë publishing *Jane Eyre* under the pen name Currer Bell, Mary Ann Evans writing as George Eliot, or Joanne Rowling writing as J. K. Rowling.

Stephanie Shirley specifically did not emulate male behavior in the running of her company. In fact, she more or less did the opposite—she looked at the reasons women were being excluded from the industry, saw them as strengths, and built a phenomenally successful company on that. What were those reasons?

First of all, there's all-out prejudice. This meant that Shirley believed there were women programmers out there who were better than some of the male ones but weren't being hired just because they were women, and especially because it was standard at the time for women to stop working when they got married or when they had children. Thus, hiring them would already put her company in a strong position. But then there was the more tangible issue of women having babies and then only wanting to work part-time, and perhaps only being able to work from home with irregular hours. Not hiring women on those grounds is now seen as something that contributes to overall structural sexism in society, but many people, including traditionalists, more hard-nosed capitalists, and some people who run small businesses, still object to this idea, because they think that anyone who wants to work short, irregular hours from home is genuinely a liability, whether they're a man or a woman, a parent or not.

This has to some extent been addressed by legislation forcing companies to disregard this fact, in the sense of allowing women to take maternity leave (and to some extent allowing men time off as well) and allowing them greater flexibility when their children are very young. However, it is worth noting that companies still find ways to get around this, and in the United States the rules are very weak compared with those in other developed countries.

Stephanie Shirley took this further: rather than just disregarding the issue, she built her company on it, specifically hiring mothers of young children. Writing code is, after all, something that even in those days did not need to be done in a specific location at specific hours (programs were written on paper and only implemented on a computer later). Shirley's view was that mothers with small children had the potential to be the most motivated and efficient of all workers, knowing that their uninterrupted time was limited but desiring, as women's liberation spread, to have more intellectual stimulation and to contribute to their family's finances. Importantly, she paid for the work done rather than paying by the hour, thus valuing efficiency over presenteeism, and perhaps playing to the strengths of mothers who knew they had limited time while their children were napping and before their husbands came home expecting dinner to be on the table.

Stephanie Shirley's story has a somewhat annoying second part: when the Sex Discrimination Act of 1975 came into force in the United Kingdom, her company was one of the first to fall afoul of the law, and she was no longer allowed to favor women in her hiring. It is a sad indictment that a new law to stop the marginalization of women was immediately used against a company that had been a champion of women from the start. However, we can learn from all of her story, including this part.

The story operates on two dimensions simultaneously. There is the dimension that is about explicit and implicit discrimination against women, including girls' schools not teaching math, women not being expected to go to university, workplace sexual harassment, and women being expected to stop working when they got married or certainly when they had children. There is the dimension that is about cultural and structural issues that made it difficult for women to thrive, especially after having children, including the culture of presenteeism and inflexible working hours.

Shirley addressed both of these issues by hiring only women, especially those with young children, and by creating a different sort of working environment in which they could thrive. In modern-day parlance, hiring more underrepresented people is about diversity, whereas creating an environment in which they can thrive is about inclusivity. In Shirley's story, and in the larger world, the two are related, but because that relationship is neither fixed nor clearly definable, I think it is more illuminating to consider them separately, to make sure we are not hiding assumptions or missing issues.

To address explicit and implicit bias may need explicit interventions. However, if we *only* address that type of bias, then we end up with more diversity and not more inclusivity. If we hire more women but maintain a work environment in which they're miserable, then they'll either quit or stay around but be miserable and thus unable to work well, which is in turn likely to perpetuate the idea that men are better at the work. There is also the danger of being accused of "sexism against men."

If, by contrast, you only address cultural issues, then you may be subscribing to the idea of gender blindness, or the idea that we should "just" treat everyone as people and value them for what they bring, regardless of gender. This would be all very well in an already egalitarian world, but we're not there yet, and this approach fails to address past or broad inequalities in society.

We really need to address both, as in this chart:

Stephanie Shirley did address both. As a result, her company was extremely successful, even by traditional capitalist measures.

Back when Shirley established her company, an extreme intervention to overcome previous gender discrimination was needed. Maybe we no longer need interventions as extreme as entire companies only hiring women, but we still need to address both issues of gender and environment. I think it will help to have a way to think about character and culture as separate from gender, and to use this to think about how to change our structures, as Stephanie Shirley did.

CHANGING THE CULTURE

Instead of giving women extra help, some of these things could be addressed by changing the entire culture of what is valued and what isn't valued. We could recognize that some skills traditionally associated with women are very important for any community, institution, or organization, such as nurturing, communication, getting the best out of people, and that these skills may go along with having sensitive emotions. Universities could value teaching, mentoring, outreach, and public engagement more than they do now, and instead of only valuing research in terms of specific results, they could also value the organizing of conferences, which is often the important catalyst for the research to get done at all.

There was an investigation in the 1990s into why male undergraduates did so much better than female ones at Oxford and Cambridge, even though the exams were anonymous and so the discrepancy couldn't really be put down to direct bias.* Perhaps surprisingly, the gender difference in history exams was even bigger than the one in math.

The finding that stuck with me was that men tended to write essays that took a strong position and argued it fiercely, and that this was highly valued. A balanced position argued from all points of view was valued less. (The report had many more facets than this.) One solution is to train women to make more one-sided arguments. But is that really beneficial to society? We could instead actually think about the value of balanced arguments, regardless of the gender of the person making them.

Rather than celebrating Emmy Noether for being a "brilliant woman mathematician," we could value collaborative

---

*N. G. McCrum, "The Academic Gender Deficit at Oxford and Cambridge," *Oxford Review of Education* 20, no. 1 (1994): 3–26.

contributions that are at the foundations of progress, not just the grand results of figureheads. Another example is Rosalind Franklin, whose contribution to Francis Crick and James Watson's work on the structure of DNA was arguably undervalued until quite recently. Like Noether, she died tragically young, and, like Noether, her work and insights were arguably crucial to a major scientific discovery. But she didn't make the big breakthrough. This is not to say that the foundational work is as critical as the big breakthrough idea, but perhaps it deserves more recognition than it has historically received. And, as it happens, such foundational work is often done by women.

What would the world be like if everyone collaborated instead of competed? I will come back to this question later, but here I'd like to give a more trivial but still indicative example. I think of it every time I'm waiting for my checked luggage to appear on the baggage carousel after a flight. When everyone crowds around right next to the conveyor belt, it is extremely difficult to see what is going on. Nobody can see their luggage until it's right in front of them, and then there's a mad scramble to grab it in time as it glides past.

If everyone stood back a few steps, however, then we'd all have a better view of our luggage as it approached, and we could step forward in time to retrieve it in a less panicked fashion, plus there would also be space for us to haul it off the carousel without bashing into people. Still, if everyone stands back, then inevitably someone will take a little step forward to get a better view than everyone else, and then someone else will step forward to see past that person, and so on, and pretty soon everyone is crowding around again and nobody can see.

There would be an uproar if I claimed that stepping forward and getting in everyone's way is typical masculine behavior and that considering the experience of other people is feminine behavior. There would be an uproar if I even claimed that whenever I experience it I see that it is men who are the first to crowd around and get in everyone's way; most people would see through my claim of it being "my experience" and consider that I was just being sexist. A divisive argument about stereotypes would then follow, and the principle of a better system of luggage retrieval would be forgotten.

This is why we need different vocabulary to talk about character traits. We need vocabulary that is not associated with gender, to give us a way of avoiding those futile divisive arguments. But we don't just need new vocabulary; we need a whole new way of thinking about the world and structuring the world. A whole new theory of people.

PART II
UNGENDERED THINKING

CHAPTER 4
A NEW DIMENSION

DURING THE AUTUMN of 2016, the Chicago Cubs were making their way toward winning that thing they won. All sports confuse me and American sports confuse me most of all, but I believe the Cubs play baseball and that the thing they won was a big deal and something they hadn't won for a long time.

During the buildup, it seemed to me that all of Chicago was going crazy about this except me. This included people who don't usually pay that much attention to sports, such as musician friends of mine and many women.

On the day that they actually played (and won) the last phase (game? match? tournament?), I was scheduled to give a talk for the nonprofit Illinois Humanities. It was in a series of events for practicing artists, consisting of a talk—not necessarily about art—and then dinner by a local chef, during which discussion could continue.

I began to wonder if I should suggest rescheduling. When we booked it we had no idea the Cubs were going to get this far, and I was worried that nobody would turn up. But people did turn up—a large number of people, in fact—and then I realized that it was a fascinating filter: it was a bunch of artists who really didn't care about a sports team winning something. My students, who at that point were all art students, felt much the same way. We all shrugged together and were a bit nonplussed by the level of excitement throughout the city.

It got me thinking about people who aren't interested in competitions, or in winning and beating other people. The types who are more interested in creating things, and in open-ended discussions that do not have clear-cut results or resolutions. It would not make much sense to refer to that as "feminine," and indeed the idea that it is somehow "unmanly" may be one of the things that puts off more (straight) men from creative arts. This is one of the many problems we can avoid if we decouple character traits from gender (aside from the sheer fact that it doesn't make sense). Describing certain men as "feminine" and certain women as "masculine" suggests that there is something wrong with them. This is something that detractors use against Michelle Obama, and some take it even further and declare not only that she is "masculine" (because she is strong both mentally and physically, exudes confidence, speaks up, and had her own career independently of her husband?) but even that she is in fact a man (because surely only a man could have those characteristics?). This is a particularly egregious conflation of gender with character.

The issue of character and gender is often brought home to me when I'm teaching. While it can be useful for me to be aware of certain statistical trends, it is crucial for me to respond to each student in front of me as they are, not as the previous statistics might suggest. However, the previous statistics do suggest things I should be on the lookout for: Girls who are convinced they are doing badly when actually they're doing quite well. Boys who think they're doing brilliantly when their proofs are actually full of holes. Girls who are too scared to speak up in class. Boys who talk a lot and unremittingly dominate the conversation in the classroom.

I see similar trends among adults when I give public lectures. In one of my public presentations I typically ask if someone will do a juggling demonstration. Usually a man's hand will shoot up and he will stride happily to the front. Sometimes there is a long silence and things get a little uncomfortable. Then I plead with the audience to help me out because otherwise I will have to do the demo myself and it might not go well. Eventually a woman might timidly put her hand up. She typically comes to the front and apologizes, saying something like "Sorry, I'm not very good, I'm really out of practice, I haven't done this for so long." And some of the timid women turn out to be much better at juggling than some of the confident men.

During the question period there are typically many men (usually white) who instantly put their hand up to ask "questions" that are often just criticisms thinly veiled as questions. Meanwhile, many women quietly ask me genuine questions privately afterward, saying they were afraid to ask in public. This trend is so marked that I have devised ways of taking questions after talks that do not involve anyone raising their hand or asking their question in public (about which more later).

However, all these trends are stereotypes, and even if they are backed up by "statistics" and "averages," the sweeping statements erase individual experiences and ignore the fact that not all men overestimate their abilities and not all women are timid and scared. As a further nuance, these characteristics might be caused by gender—because of the way society treats men and women differently, they may well develop different levels of confidence or timidity in public—but this does not mean these are inherently male or female characteristics. So we would benefit from ungendered terminology to talk about them, instead of saying that confidence is "masculine."

Ungendered terminology could then also help us think about how to encourage different characteristics that we wish to value and nurture in both men and women. In *Crossing* Deirdre McCloskey writes of her new experiences when she transitions to being a woman, and among other things finds that women will help her. On the other hand, as she writes, "female-to-male gender crossers must face the unhappy fact that American men don't help each other. The theory of American maleness is that your special woman takes care of you when you're sick, but aside from that you are supposed to do everything alone. Help among men is shameful, because it shows incompetence. Among women help is the point, because it shows love, 'love' in its full sense: care, sympathy, providing for need."

There is no reason that this needs to be a male/female split in the end, other than a deeply ingrained cultural one. I believe that if we adopt ungendered language to talk about these things, then we can escape from those cultural restrictions and everyone can be freed to benefit from the love and help of others.

Gradually, through all these experiences and thought processes, I came to realize that we need to think about new concepts, or think about old concepts in a new way. And I realized that new terminology would be a crucial key to thinking about these new concepts effectively.

We need new, ungendered language in order to separate character traits from gender and have less divisive conversations in which people don't have to get defensive about "not all men" or "not all women" being a certain way. Because indeed men are not all the same, and neither are women. Not all men are aggressive, competitive, risk-taking, and unempathetic. And even those who do behave in those ways might only do so because of social pressure and the idea, perpetuated by social norms, that this is how to be successful in society. When certain behavior is rewarded by society many people will strive to behave in those ways even if at some deeper level it makes them unhappy.

Men and women currently suffer in opposite ways from these pressures: both may be unhappy with traditionally masculine behavior, but if women decide not to emulate it, they are at least accepted as feminine, whereas men may be criticized for a lack of "masculinity." For example, if a woman earns less than a male partner (or is younger or less educated), it does not raise eyebrows nearly as much as if a man earns less than a female partner (or is younger or less educated). By contrast, if a woman takes on those "masculine" characteristics, she may well be more successful, but she may also be criticized because those same characteristics are seen as negative in a woman; where a man is ambitious and confident, a woman is pushy and aggressive. In a fairer society the pressures on men and women would be the same: whatever it is we value and encourage, we should value and encourage it from all people regardless of gender. This is not about being "gender-blind" overall: we still need to see gender in order to fight injustice that is directly gender-based. It is about learning how to stop perpetuating subtler forms of systemic bias that favor men by favoring—for no good reason—characteristics associated with men.

My experience of gender imbalance and gender norms has changed dramatically now that I teach math at the School of the Art Institute of Chicago (SAIC). This is an "art school," in the parlance of American education—that is to say, a university in which all degrees are in aspects of art: fine art, sculpture, architecture, design, fashion, photography, animation, and so on. I previously taught traditional math students at traditionally academic universities, and my classes were often around 90 percent male. By contrast, only about 40 percent of the art students at SAIC identify as male, and, as you might expect, the male students do not conform to really any of the traditional "masculine" character types. Some of them, especially the more straight-passing white ones, do carry with them some of the benefits of having grown up with society on their side, such as easier confidence when speaking up in group situations, but not all of them have benefited in this way, and many of them are sensitive enough to the needs and voices of others that they tone this down in order to make sure others have a chance to participate.

It would be divisive, insulting, and I believe obstructive to say that these men are less "masculine" than their aggressively competitive, probably more financially successful peers in the non-art world. We need a way to talk about character traits without reference to gender, and new axes along which to measure them.

Mathematically speaking, if we have two things that are not equal, we could make them equal by making the lesser one greater or by making the greater one less, or by a combination of both. However, there is a completely different way we could do it, which is by evaluating the two things on a new dimension entirely. We wouldn't necessarily make them equal in the end, but we would at least have made sure neither of them was affected by the original inequality.

To take an example, if two people were interviewing for a job as a typist, we could equalize their heights by getting the shorter one to stand on a box. Or we could observe that height is irrelevant to working as a typist, so we should evaluate how well they type instead.

This sounds silly, but getting shorter people to stand on a box is essentially what we are doing all the time we wonder about how and why men and women are different and what we should do about it. Instead, we could just understand people's relevant characteristics, which much of the time have nothing to do with their gender, instead of trying to use gender as a proxy for character. This doesn't mean that we are making men and women equal in a strict sense. Men and women are in fact not the same. What it means is that we are making sure that being a man or a woman doesn't have an effect unless it really has to (in a gynecologist's office, for example). This is the idea of finding an entirely new dimension on which to think about things, a dimension that is separate from the dimension of gender.

A new dimension is a way of escaping the old dimensions. A train is constrained by having to run on a track. That is one-dimensional (except where it branches), as the train can only go forward and backward, which is just the positive and negative of the same direction. If someone wants to block the train, all they have to do is block it in front and behind. But you could escape by invoking a second dimension: you could open a door and jump out. At this point you would be moving in a left-right direction that the train can't do.

Dimensions are directions, but they can be directions of ideas rather than of space. If we are thinking about choosing a restaurant for dinner, we might evaluate our choices according to price, cuisine, ambiance, distance from home, and so on. Each of these is an abstract dimension, a dimension of ideas.

When considering dimensions of ideas, a new dimension still helps us to escape a trap; it's just that we're now escaping a trap of ideas. Perhaps this is the image that is being evoked if we use the now awfully clichéd imagery of "thinking outside the box": we are using a new abstract dimension to escape limited low-dimensional thinking.

Sometimes the higher dimensions were there already. Life is complicated and we have to simplify it in order to understand it; one way to do that is to depict it in lower dimensions, like taking a picture, which is a two-dimensional depiction of our three-dimensional world. If we try to reconstruct a scene from a photo, we're not exactly "creating" a third dimension;

rather, we're trying to rediscover the dimension that was eliminated by the photo. If we're dealing with dimensions of ideas, using a higher dimension consists of stopping the urge to squash a direction of thinking. That sounds like it will make things more complicated, but sometimes it's only a short-term complication and can simplify things in the long run.

In Chapter 1 we talked about how math uses a new dimension to escape the constraint of being unable to take square roots of negative numbers. Mathematicians came up with a new dimension of number, which they called "imaginary," and we represent it as a second dimension, usually vertical where the ordinary number line is horizontal across the page. The basic unit of this new direction is called i (for imaginary), where i is declared to be a square root of -1. Then the two-dimensional space that we can access by combining this direction with the old "real number" direction is called the complex plane, and the points represent numbers called complex numbers, which are a blend of real numbers and imaginary numbers. Here is a picture of the complex plane and some complex number positions on it:

Although in some sense imaginary numbers were just "imagined" by mathematicians, it turns out that this extra dimension helps us to see certain things more clearly. For example, in the real numbers the number 1 has two square roots, 1 and –1, because $1^2 = 1$ but also $(-1)^2 = 1$. However, 1 only has one cube root (which is 1), which is a bit unsatisfactory to mathematicians—why aren't there three, since we're now dealing with a power of three? In the complex numbers we can show that there actually are three cube roots of 1,* and moreover, if we plot them on the complex plane, we see that they are equally spaced around a circle of radius 1:

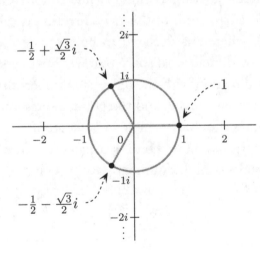

---

*We can do some multiplying and show that

$$\left(-\frac{1}{2} + \frac{\sqrt{3}}{2}i\right)^3 = 1 \quad \text{and} \quad \left(-\frac{1}{2} - \frac{\sqrt{3}}{2}i\right)^3 = 1.$$

This is a calculation I would call tedious but not difficult.

This geometric picture helps us to generalize to n-fold roots of 1, the complex numbers x such that $x^n = 1$; these are also equally spaced around a circle of radius 1. Here is the picture for the eighth roots:

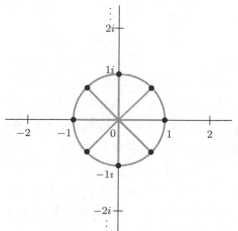

The picture and pattern are much more vivid in two dimensions; on the one-dimensional line of real numbers we can't really see anything.*

This is one example where using more dimensions is more complicated but also gives clarity and nuance, and it is typical of what using higher dimensions does for us. The complication means that we will always be tempted to revert to lower dimensions for convenience, just as it is tempting to take a photo rather than try to remember all the details about a scene. This is fine as long as we recall that there are many

---

*The whole picture is even clearer when using a polar coordinate expression for complex numbers (expressing points in the plane by a radius and angle) rather than the one we have here, which essentially uses Cartesian coordinates (expressing points in the plane by an x-coordinate and y-coordinate).

different possible angles from which to take a photo, and some angles will miss crucial details that will be captured by others. Being able to see things from different angles can be a decent substitute for actually working with the higher dimensions.

A great deal of my own research in higher-dimensional category theory involves moving between higher-dimensional structures and their lower-dimensional "footprints." The questions go in both directions: What sort of lower-dimensional footprints will a higher-dimensional structure leave, and what can we deduce about higher-dimensional structures from their footprints? Often the footprints lead us astray and give us mistaken impressions about the higher dimensions, like the two-dimensional shadow in Chapter 1, which looked like a foot but turned out to be my hand.

I love the novel *Miss Smilla's Feeling for Snow* by Peter Høeg, and one detail that has always stayed with me is that Miss Smilla understands footprints in snow so well that she is able to reconstruct a child's last moments from the footprints he has left behind. She grew up playing a game in which children jumped around in the snow while someone closed their eyes, and that person was then challenged to work out what sort of jumps they did to produce those footprints. Sometimes this is what my work feels like.

Thinking about gender sometimes feels like this to me too: that we have taken a complex higher-dimensional structure of people's characters and encapsulated it in this one dimension in a flawed and reductive way, and our attempts to reconstruct the higher dimensions are not going very well. We need to do better.

It may seem that I'm simultaneously talking about finding a "new" dimension and rediscovering one that we have squashed out of our consciousness. In math the question of whether we invent things or discover something that was already there is a bit moot. Personally, when I'm doing new research I feel like I'm *discovering* concepts that were already there, but *inventing* ways to talk about them. The latter aspect is important. Sometimes people say math is "just a language," as if language is a small thing. It's not.

Language is important. It can help us clarify our thoughts about something, even if we were already thinking about that thing before we had a word for it. Having a word for a concept is a way of making it more convenient to carry around in our brains. If you're not exhausted from carrying something around, you can carry more things, take them further, and do more with them.

This is how we make new concepts in mathematics. Arguably the concept itself isn't new—it is there whether or not we give it a word. But giving it a word can open up whole new trains of thought, and can enable us to build on the idea and carry it further.

One basic example of this is multiplication. Multiplication of numbers can be thought of as just repeated addition, so 4×2 is

$$2 + 2 + 2 + 2.$$

In a way, multiplication is not a new concept and you don't really need a word for it, because it could be expressed as addition. But various new things become possible if you think of it as a new concept. First of all, you can do much more complicated versions, like 55 × 65, which would take a long time to do as repeated addition. But also we can now build a new concept from multiplication by repeating it in turn. 4 × 4 × 4 gives us exponentiation, written 4^3. If we had to express this using only addition, we'd have to do

$$(4 + 4 + 4 + 4) + (4 + 4 + 4 + 4)$$
$$+ (4 + 4 + 4 + 4) + (4 + 4 + 4 + 4)$$

which is rather arduous. Furthermore, we can think about versions of multiplication that aren't so much like repeated addition at all. There's a concept of multiplying shapes together, multiplying patterns together, multiplying symmetries together. This all began by thinking of repeated addition as a concept in its own right.

In this chapter I'm going to suggest new vocabulary to encapsulate some concepts that we may well have been thinking of all along. My aim is to help free us from all the problems we've discussed arising from our conflation of character with gender.

NEW LANGUAGE

The word "feminism" is gendered and thus divisive in many situations. I once bought *We Should All Be Feminists* by Chimamanda Ngozi Adichie and *Why I Am Not a Feminist* by Jessa Crispin at the same time in an airport book shop. I thought this was quite a funny combination and made a comment about it to the checkout guy, but he gave no flicker of recognition whatsoever. I suppose I was just part of his job; I'm certain that if I had done this at my local family-run bookshop, Sandmeyer's, we'd have launched into an interesting discussion about it.

As it turns out, those two books do not exactly contradict each other; they just use different definitions of feminism. I have discussed the fact that some of the issues facing women are directly because they are women, that some are only indirectly so, and that there are many issues people face that are nothing to do with whether or not they are women. The divisive arguments over these competing issues—and the fact that they are in competition at all—act as a de facto tool serving the people who are already in power and helping them to keep power.

Women who feel kept out of power are developing language to help them try to change that balance. That language may well serve to motivate and unite some women and get them focused on what they want to achieve, but at the same time it too often alienates other people: men who might otherwise be allies, or more disadvantaged women who feel overlooked by the women who are themselves in positions of power among women.

One such sharply divisive word is the verb "mansplain," coined by Rebecca Solnit. Women around the world breathed a collective sigh of relief and recognition when we were finally given a word for this thing that we feel happens to us often, where previously we couldn't quite put our finger on it or be sure we weren't just imagining it. I have written elsewhere about straw-person arguments against this concept.* A few men have declared that they are unfollowing me on Twitter just because I use the word (I don't miss them). Some people declare that "women mansplain too," showing that they have misunderstood what the concept is. To reiterate: Mansplaining is not just when a man explains something to a woman in a patronizing way. It is when they do it despite there being clear signs that the woman already knows it and possibly knows it better than they do, which fits into a pattern that many women experience across their whole lives, of society generally assuming that men are more knowledgeable than women. Thus women widely experience men discounting their knowledge and expertise, talking over them, claiming their ideas as their own, or even repeating what they've said immediately after they said it, as if it were a new idea. Society doesn't have a general widespread assumption that men are not experts in their field (though it does happen in some specific areas such as childcare), and so women by definition cannot engage in mansplaining. It's a bit like picking on someone smaller than you: if you're smaller than the other person, you cannot be described as picking on someone smaller than you.

---

*Straw-person argument: when you invent a different point that nobody is making but that is easier to knock down, and then knock it down. But you haven't defeated anyone's argument, because nobody was making that argument. Previously known as "straw-man argument." See Eugenia Cheng, *The Art of Logic in an Illogical World* (New York: Basic Books, 2018).

However, I acknowledge that this idea really winds some men up. Whether that matters or not is a separate question; either way, it's an example of how divisive language can be.

Another word that women and feminists are increasingly throwing out is "patriarchy," to acknowledge and draw attention to the fact that in our society as it currently stands, power is stacked in favor of men and against women. This word also riles up some men, often when they individually don't feel that they have much power in society. But the point is that far more men than women do hold power, and the *structures* of society favor men, so this skewing of the scales is deeply embedded. Some men may particularly feel that they have not benefited from the so-called patriarchy—for example, gay men and poor men. One black male professor has told me that he thinks all women, including black women, are better off than black men in terms of the support they receive in life and the outcomes they experience.

For these reasons I try to qualify references to "patriarchy" by calling it the "straight white patriarchy" or the "straight white rich patriarchy" or the "straight white rich cis-patriarchy," but this leads me to wonder how many more adjectives we will have to attach to this situation. I do believe that it is important for us to take all these identities into account when considering the fairness of the world, even though some people, usually straight white men, roll their eyes at "identity politics." But importantly I think we also need a way to think about relevant characteristics of people *separately* from the whole identity discussion. We need to stop blurring the issues. When someone's identity is relevant because of prejudiced attitudes of people around them or inequality embedded in society, then that is a genuine problem, but when we're thinking about characteristics or abilities, those things do not need to have anything to do with gender or other identities. Whether or not they correlate with gender is not relevant, because correlation is not causation and does not completely determine what will be true about someone.

I have already talked about several of my experiences of women who do not fit the traditional "feminine" type, including some of the professors at the women's college I was at and the Five Billion Dollar Woman at the party. I have also seen character types that might be associated with femininity but which include many men, such as all the artists not interested in sports, as I described at the beginning of this chapter.

To address our need to separate gender from character, I took a mathematical approach, not shying away from reorganizing existing thoughts into new packages and giving them names. In math when we name new concepts we sometimes use old words in new ways, if we are particularly seeking to make connections with previously known concepts. But sometimes we are trying to clear our minds and get rid of prior associations, and that's when we invent new words. That's also why we need new words here, to break our prior associations with gender and clear our minds.

I brainstormed for a year or two on and off with my dear, empathetic, emotionally sensitive, supportive friend Gregory, and came up with the following words: "ingressive" and "congressive." The etymological idea is that "ingressive" is about going into things, and "congressive" is about bringing things together.

These are descriptive words rather than prescriptive ones. They just describe behavior, attitudes, or situations. Using descriptive words can liberate people of all genders, and in particular, men can be liberated from gender pressures as well. There might be a tendency to immediately constrain ourselves according to these new criteria, and find ways to assess who is born ingressive and who is born congressive, but that is absolutely not the idea, as we'll see. Perhaps people feel drawn to putting themselves in categories just as much as they resist being put in them.

It is possible that these behaviors correlate with men or women in some ways, but that is a sort of afterthought, not a driving principle, and in any case it is not an innate quality but one that is learned through society's conditioning and so can be changed. It's possible that treating people fairly according to these characteristics will lead to changed outcomes in terms of gender equality as well, but the main point is that it will lead to changed outcomes in terms of *human* equality. We will still need to address sexism, racism, homophobia, transphobia, income inequality, and all forms of prejudice, bigotry, and exclusion, but we will have greater clarity if we have better language to keep these issues separate from character traits. I will argue that these concepts and this terminology open dialogue and build bridges across identities rather than making our arguments more and more divisive and splintering us into more and more distinct and competing tribes.

With this ungendered language I can now rethink many of my past experiences and case studies with greater clarity. I can say that the Five Billion Dollar Woman was very ingressive, and this makes more sense than trying to say that she was very "masculine" in her behavior. Emmy Noether's work and the work of the Bletchley women was congressive. Stephanie Shirley ran her business in a way that not only addressed discrimination against women directly but also created a congressive environment in which congressive people could thrive. The books about women who are "headstrong" or "badass" are trying to portray women as great because they can be ingressive. Sheryl Sandberg seems to be asking women to be more ingressive in order to be successful in an ingressive world. This certainly is one way to be successful, but we are going to examine other ways to be successful that we can approach more clearly now that we have this new terminology to help us.

Before we do that, it is important to note that I am not claiming that this is a clean dichotomy. People aren't just ingressive or congressive. They can be somewhat one and somewhat the other under different circumstances. It's more of a two-dimensional plane, and we can be anywhere on it,

ingressive

congressive

and indeed move around on it across time and in different situations.

Importantly, these are not fixed characteristics but just types of behavior, and like all behaviors, they can be learned. At a certain point I realized that I had spent most of my life learning to be more ingressive to be successful in an ingressive world, perhaps as Sheryl Sandberg suggests. But I didn't like myself that way. In my new multifaceted career I have been unlearning that ingressive behavior and finding that I can be successful while remaining true to my congressive self. Having the language to express it that way has helped me work out how to go about doing it.

Here are some fuller definitions or characterizations of these words.

ingressive: focusing on oneself over society and community, imposing on people more than taking others into account, emphasizing independence and individualism, more competitive and adversarial than collaborative, tending toward selective or single-track thought processes

congressive: focusing on society and community over self, taking others into account more than imposing on them, emphasizing interdependence and interconnectedness, more collaborative and cooperative than competitive, tending toward circumspect thought processes

This is not a clean dichotomy even though some aspects sound like exact opposites of each other. It's not a classification of people into two camps. It's a way to evaluate behavior in a flexible and dynamic way to reflect the fact that people are flexible and dynamic day by day and also over the course of their lives, not fixed and rigid.

RELATIONSHIPS WITH EXISTING IDEAS

Although people are flexible and dynamic, the urge to separate their behavior into (binary) gender categories is strong, and ideas based on those distinctions aren't completely useless. Those distinctions can be useful *to some extent*, but I believe the usefulness comes with many unhelpful side effects and that a new, ungendered approach can preserve the usefulness while eliminating the side effects.

Indeed, most of what I read about gender issues seems to me to be really about ingressive and congressive issues, just without the terminology. And if it continues to relate the issues to gender, then it risks alienating those who do not fit the behavior descriptions. We also risk detracting from the point by causing arguments about whether a behavior really includes all men or all women and whether it's patronizing to simplify men and women in this way and simplistic to make such a dichotomy.

Men Are from Mars, Women Are from Venus by John Gray is a famous and divisive example, but I actually found it rather useful. I didn't take it to mean that *all* men behave in the "Mars"-type ways, but it helped me to recognize when someone (of any gender) was behaving in one way while I was behaving in the other, and thus helped me to

communicate better and resolve situations that might otherwise have become ever more antagonistic. It helped me
understand things like why I am resistant to advice when I
tell someone about a problem in my life (because I am seeking validation, not a solution—in a Venus-like manner). I
take the book to be really saying, "Some people, in some
situations, are from Mars, and others, in other situations,
are from Venus, and it's often men who are one and women
who are the other but not necessarily, and you might be
both at different times in different situations." That is a
somewhat less catchy title.

Somehow the imagery is much more vivid and arresting when it involves a very distinct dichotomy between two
completely different things, but our need for something vivid
and arresting can get in the way of a nuanced understanding.
Perhaps it's because we've been influenced to be too ingressive that ingressive dichotomies are necessary to attract our
attention?

With the terminology of "ingressive" and "congressive"
we have a chance of avoiding those issues because we are
focusing mainly on the characteristics, not on the genders.
It is a separate question to wonder whether men are more
ingressive and women are more congressive, and if so, how
much of that is learned. But that is not the main question we
will focus on. If we focus on the characteristics instead of the
genders, we have more chance of seeing those characteristics
as flexible and dynamic, both across people's lifetimes and
day by day or minute by minute in different situations. Furthermore, because this theory is somewhat abstract, it offers

the possibility of shedding light on many different situations at once, addressing different parts of life, looking at situations at different scales, and unifying many existing ideas about contrasting character types.

The Mars/Venus imagery largely focuses interactions in personal relationships, but there is also language aimed at professional interactions. *Work Like a Woman* is the memoir by Mary Portas, a British retail and brand communication specialist who is also known as "Mary Queen of Shops" and "Queen of the High Street." Ostensibly she is writing about how she became successful and high-powered as a woman in a male-dominated business world. She also seems to me to be talking about ingressive versus congressive styles of business and leadership, though she doesn't use those words. She talks about the "alpha" and "macho" culture of traditional business, which are good words but are still very strongly associated with men. She also talks about "vertical" as opposed to "circular" ambition and career paths. ("Circular" is not supposed to invoke the idea of going around and around in futile circles but rather give the impression of emanation and non-hierarchy.) I am slightly afraid that these words have too obvious a connection with gender; it's only a small mental shift from vertical and circular to phallic and yonic.

Jessa Crispin discusses related ideas in *Why I Am Not a Feminist*. She writes, "The era of domination has to be replaced with an era of collaboration, not segmentation." She calls on feminists to "create a world of cooperation and fraternity," curiously using a male-gendered word at the last moment while essentially rousing us to create a more congressive world.

HOW NEW TERMINOLOGY HELPS US

Evidently I think that many insightful people have been writing about ingressive and congressive issues for a long time, just without the help of ungendered terminology. Here is how that terminology can work for people of all genders as a new way of understanding the world and perhaps even changing it for the better. These ideas come from my field of research, category theory, which has opened up huge new vistas of mathematical thought by giving subtly new points of view. One of the points of view has to do with how we describe or characterize things.

There are often two ways to describe something: by some intrinsic characteristics or by the role it plays in a certain context. We could describe Barack Obama as male, a little over six feet tall, athletic, short graying hair, or we could characterize him as the forty-fourth president of the United States or the first nonwhite president. Each of those descriptions is true, but each accomplishes different things and draws our attention to different things.

In category theory, abstract mathematicians are very interested in describing things by the role they play in a context rather than by their intrinsic characteristics. It is in a sense a more versatile way of thinking, as it means we are focusing on what is relevant to a situation. Sometimes something like hair color might be relevant to playing a role—for example, if you are trying to be someone's body double. But in many contexts it is not relevant, such as if you're doing scientific research in a lab, going up in a spaceship, or doing gymnastics—although all too often journalists still feel the need to describe women's hair when writing about them doing these things. Category theory says we should focus on what is relevant in any given context, which is determined by the roles that things play in that context, regardless of what their intrinsic characteristics are.

This is the approach I take to describing character types according to the new terms "ingressive" and "congressive." It is about the roles that people and behaviors play in different aspects of society, not about people's gender, race, sexual orientation, wealth, or any other intrinsic measure.

I think this approach is fruitful and important in its own right whether or not it leads to greater gender equality, although I believe it will lead to that, as well as to greater equality along other dimensions. I think that if we remove gender and other intrinsic identities from our arguments, our arguments can become less divisive and more productive. That means focusing on identity when it is relevant (which it often is, especially when we're thinking about prejudice) and not when it isn't. This amounts to separating out feminist arguments into the parts that really are about gender and the parts that are really about character, which might be more implicit or structural. We are separating out character and gender as independent variables, as we described in Chapter 3, instead of assuming that there is some formula relating the two. The existing approaches consist of continually trying to find a formula for character based on gender, and continually finding that any formula would be inaccurate, divisive, and dehumanizing.

The best formula, really, is to admit that there isn't one. In reality the variables are probably neither fully dependent nor fully independent, so we do have a choice about how we treat them. If we treat them as dependent, then we fall into all the traps of divisive arguments and prescriptive gender associations. If we treat them as independent, all the arguments become stronger, because we no longer need to pit genders against each other; instead, we ask what qualities of humans we are genuinely going to value in society. We can then work to nurture, value, and reward those qualities in everyone of all genders. This is an important way in which individuals can influence structures, which can then influence individuals in turn.

This is a cycle that is too often vicious, but we can work to make it virtuous instead. Sometimes we are obstructed in this by people denying that structural pressures exist. If an individual is personally prejudiced or oppressive toward another, that is one thing. But even if no individual is actively behaving in that way, it is possible for society to have that overall effect on people, because of how society is structured. This difference between individual bias and structural bias is the cause of a great deal of antagonism and misunderstanding of identity politics.

Here is a diagram depicting some structural reasons that men continue to be more successful and powerful in the world, to help us understand what we are currently doing about it and what we could do instead. I am writing "Y" for any general characteristics that are traditionally associated with men such as self-confidence, ambition, competitiveness, and risk-taking, but eventually we can consider Y to be general ingressivity.

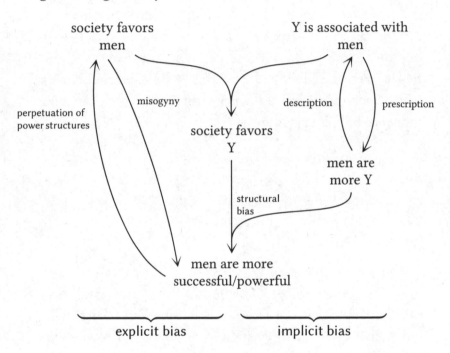

This diagram divides broadly into two parts. The left-hand side is explicit bias—that is, when society perpetuates the structural power of men through direct discrimination against women. This is the part that definitely needs to be addressed in a way that remains conscious of gender. However, if we're not careful, this comes out as "sexism toward men," although by some contemporary definitions of sexism it only counts as sexism if it's by the group that holds power (men) against the group that doesn't (women). In any case, it is divisive because it's still based on gender, and some people think this is now "unfair" toward men in return.

I would call the right-hand side an aspect of implicit bias, in which society favors men not just because they're men per se but because of the way society favors certain characteristics that happen to be associated with men, perhaps all unconsciously. I believe we are currently addressing the character bias in terms of gender, and this is causing other problems, while also not fixing the old ones.

If we just address the fact that men are more likely to have the characteristic in question ("men are more Y"), this results in pseudo-feminism, or "leaning in," the idea that women should learn those characteristics and "become more like men" to be successful. Another approach tries to fix the fact that these characteristics are associated with men ("Y is associated with men") by doing what gets referred to as "redefining masculinity" or persuading men to take on other characteristics that are more associated with women. Whether we are asking women to be more like men or men to be more like women, we are still basing our ideas on gender and remaining stuck in that one dimension.

But rather than addressing just the two sides of the diagram, there is a third way: the middle. We could address the fact that society favors those characteristics at all. This is the part that should not be based on gender. It is currently related to gender, but that is arguably for historical reasons rather than essential ones. I am proposing that we address this issue separately along our new dimension, which will detach it from both sides—that is, break the arrows that join it to the left and the right. When we break those arrows I believe we find that there is no particular good reason for society to favor the characteristics Y at all, and we are then liberated to think about what characteristics we would actually like society to favor.

The new language helps us with this liberation. Instead of saying "characteristics Y," we can say "ingressive characteristics." I argue that ingressive characteristics are associated with men, and men are currently more ingressive. Moreover, society favors ingressive behavior, with the result that men are more successful and powerful. But I will also argue that *congressive* behavior is actually better for society, so we should favor congressive behavior instead. This will automatically change the gender balance of power in society without us having to take action along the gendered dimension.

How can we become more congressive, as a society and as individuals? Currently the most obvious way to be successful in our ingressive world involves becoming as ingressive as possible in order to be "successful" according to those structures. But instead we can find congressive ways to be successful despite the ingressive structures of the world. Or we can change the entire structure to make the structure congressive. This second part of the book is about these possibilities for building a more congressive society. First, in Chapter 5, I will describe ways in which I think society currently favors ingressive behavior, for no particular good reason.

CHAPTER 5
STRUCTURES AND SOCIETY

OUR CURRENT SOCIETY broadly favors men, and I believe that one contributing factor is that society as it stands is largely ingressive. It does not mean that every individual in it is personally ingressive any more than it means that every individual is personally misogynistic. But I think it has led to general pressure to be as ingressive as possible in order to succeed, and that this has a knock-on effect on gender imbalances. In this chapter I'm going to explore how these ingressive structures are being perpetuated, ways in which they're bad for us, and what we could do instead.

As I am a mathematician, my main experience of building environments and structures, and nurturing people to flourish, is in math. I also think math and science are key areas that are presented ingressively when they really have very congressive facets.

I have a particularly curious experience of this because of my unusual teaching experience: I taught traditional math in traditional ingressive universities for some years before switching to teaching math to art students at the School of the Art Institute of Chicago. As I mentioned earlier, in the first part of my career I was teaching students who had done very well at school math and who had enjoyed it enough to continue in college, and the classes were very male-dominated. I switched to teaching students who had typically not done well at high school math and often hated it so much that they fled from it as fast as possible as soon as they were allowed, and the gender balance flipped.

Now that we know to go further back to more basic principles than gender, I can theorize that these different gender imbalances are related to ingressive tendencies: that we teach and present math ingressively, that this appeals more to ingressive people, and that this in turn means that it appeals more to men and puts more women off, although—as usual—if we talk about it via gender, we have to say "not all men" and "not all women." This also ties into my personal story, in which I made myself as ingressive as possible in order to be successful in an academic math career before becoming unhappy with that persona and trying to become more congressive, whereupon I discovered I didn't like that career anymore.

In teaching my wonderful, intelligent, thoughtful, self-aware, often math-phobic art students, I have taken the opportunity to learn from them about their past math experiences—to find out what put them off math in the past and, crucially, to develop ways to "put them back on." The short story is that they were put off math by its ingressive nature and the ingressive way it is taught, and they are put back on it when I show them that it is actually very congressive and when I present it in a congressive way. And I do believe that math can be congressive and that education should be congressive.

However, as with so many aspects of society, it is very deeply ingrained that math and education are ingressive, but I think this is yet another cycle of influences—it's ingressive because the rest of society is ingressive. Fortunately, the School of the Art Institute is much more congressive than all the other placcs I've taught, in the trust it places in its teachers, the autonomy I am given, and the fact that I seem to be given liberating responsibility rather than oppressive accountability. So it has been an ongoing opportunity for me to move away from old ingressive teaching styles. I have been developing congressive ways to teach math, and to develop a little utopia of a congressive society in my classroom.

My teaching is now unrecognizable compared with when I first started teaching at Cambridge University about twenty years ago. I was never exceedingly ingressive, but I was fitting in with an extremely ingressive environment, and I taught in a very traditional way, as everyone else did: by standing at the front and writing a huge quantity of notes on a blackboard. The (mostly male) students frantically copied them down, tried to understand them later, and then took a grueling exam at the end of the year.

Now I do almost exactly the opposite. Having the un-gendered terminology has helped me to change this, because without it I might have been stuck over the question of whether I was trying to teach math that would "appeal more to girls," which sounds (and is) silly. There is a great deal of formal research and literature on different approaches to teaching, and I am not claiming that my methods are new;

indeed, most of it is inspired by the best teachers I've learned from in my life, either as a student or at conferences. The thing I want to stress is how much it has helped me to have the terminology with which to unify these various concepts in my head and also with my students. It means I can always ask myself, "What is a more congressive way of doing this?" and when I think I didn't do something optimally, I can often recognize that I was accidentally slightly too ingressive. I can recognize if a student is being a little ingressive and then realize that what I need to do is neutralize that and encourage them to be more congressive, with the result that everyone learns more, including them. Many educators work hard to make more congressive classrooms, whether or not they use the terminology. However, many classrooms are not very congressive, especially in universities and especially in math, because of basic assumptions about the ingressive nature of math.

I'm going to describe how I have moved to congressive math teaching and discuss some ways in which I think it is misguided to think that math is inherently ingressive. I will then look at how this principle of overvaluing ingressivity isn't just confined to math but is pervasive in education beyond just math, and even beyond that to society at large. Education could be seen as the root of all these ingressive problems, but as always, it's a cycle: education is ingressive because society is ingressive, and society is ingressive because we are all educated to be ingressive.

In *No Contest*, Alfie Kohn characterizes competition as coming from situations where resources are scarce. But education involves a resource that can never be scarce: one person having knowledge and wisdom does not prevent someone else from having it. It might be scarce in the sense that not many people have it, especially when it comes to very specialized knowledge, but the whole point of education should be to share knowledge and wisdom with the next generation and thus ensure that it keeps growing. So the fact that we make education competitive is at worst contradictory and at best a choice that we should acknowledge and question. Ingressive behavior is about more than competition, but this is a good place to start thinking about the assumptions embedded in our system.

Assumptions that go unquestioned stop us from thinking about different points of view. A typical quip about mathematics is that "one plus one just *does* equal two." However, this is not exactly true. There are situations where one plus one is zero, such as if you say "I'm not not hungry," in which case you're really saying you are hungry: one "not" plus one "not" is zero nots. Or if you rotate something by 180 degrees and then rotate it by 180 degrees again, it's like doing it zero times.

There are also situations in which one plus one can be one. If you add a pile of sand to another pile of sand, you still have one pile of sand, albeit a bigger one. If you mix one color with one color, you get one color.

If we focus too much attention on the supposed "absolute truth" of basic arithmetic, we risk getting into a mindset where we assume not only that adding things together makes more of them but also the flipside: that using them or sharing them will make less. This is true of things like money, food, and time, but there are other types of resources that are not depleted by use. Knowledge is one such non-expendable resource, as are wisdom, curiosity, flexibility, love, joy. Not only are these not depleted by use or made smaller by sharing, but in fact they are probably depleted by non-use and increased by sharing. If you learn something but then don't use it for a long time, then you are quite likely to forget it, whereas every time you use it you strengthen your knowledge. And one of the best ways to strengthen knowledge and deepen understanding is to share by teaching. Giving love, in a healthy situation, strengthens it and generates more love.

Education is a good place to investigate issues of competition and more general ingressivity because it can be examined at many levels—at the broad level of the entire educational system, but also at more zoomed-in levels of individual schools and classrooms, classes, activities, or methods. Every classroom is a small microcosm of society, and teachers can influence those small microcosms very quickly and effectively, even if they can't affect the whole school or education system. It is also something that is a fairly universal experience, as we have all had some experience of education. Using the new ungendered terminology can help us understand that experience in more productive ways.

The first way in which the terminology helps me is that I start every course by discussing the theory of ingressive and congressive behavior with my students and making it clear that I value congressive behavior. In the classroom this means collaboration, contribution to the group, depth rather than speed, and curiosity rather than knowledge. Typically my art students are delighted by this and rapidly embrace it.

I also try to make the classroom as non-hierarchical as possible. The class is not about my power over students, it is about the process of mathematics. I think it's important not to set myself up as imparting knowledge by authority; rather, I am showing students what the processes of mathematics are, as it is those processes, not my authority, that determine truth and value in math.

Mathematician and educator Professor David Kung has a thought-provoking TEDx talk in which he points out that if we teach students to associate knowledge with authority rather than processes, then they will become adults who continue to do that and who will thus believe whomever they see as an authority figure, rather than the important processes of logic, evidence, and reason.[*] Essentially that is an ingressive approach to knowledge rather than a congressive one, and keeping that difference in mind can guide all our choices in how we set up structures in the classroom and beyond.

To help move away from a hierarchical structure, I set up our desks in a U shape rather than the traditional rows with the "lecturer" at the front. I also spend as little time as possible standing at the front doing traditional lecturing, but instead do as much exploration and discussion as possible. I nurture, encourage, and reward congressive behavior such as curiosity, open-mindedness, and collaboration, not ingressive behavior such as showing off, posturing, or belittling others. This also allows congressive people to reach their full potential without having to mimic ingressive behavior.

---

[*] David Kung, "Math for Informed Citizens," TEDxGreatMills, August 1, 2019.

Unfortunately, many traditional classrooms (especially in college) are ingressive, with all students—and the teacher—trying to show how brilliant they are, often by making others feel stupid. It is often boys doing it to girls and men doing it to women, but not always; it is ingressive people doing it to congressive people. A congressive classroom is instead one in which nobody is trying to win but everyone is trying to make sure the group is making progress and learning things.

I give students projects to explore and investigate, where there is no right and wrong answer but a "low floor and high ceiling"—that is, a low barrier to entry and no real limit to what you can learn from it. I get them to collaborate rather than compete, with the idea that the whole class, between them, will contribute to the collective discovery process, nudged along by the teacher.

This expands to the possibilities for congressive outreach as well. Making math into a competition or a game is thought to make it more "fun," but it only makes it fun if you enjoy competitions and games. This risks putting off congressive people (including me). However, I love crafts and almost always prefer a math craft to a math game. I like math fairs, where there are various tables of mathematical activities and children can spend as much or as little time as they like at each, with no particular aim. I am lucky that I was not entered in any math competitions when I was a teenager, as I would have hated it and possibly concluded that I was not cut out to be a mathematician. I worry that other congressive people have the same experience and that this leads to us shutting out congressive people from mathematics—and, more broadly, from science. Having this ungendered language, we can think more clearly about designing more congressive activities to appeal to more congressive people, without having to resort to saying silly things like "activities to appeal to girls." It may also help to explain why girls are so underrepresented in International Mathematical Olympiad teams, as we discussed in Chapter 3. If girls are being put off from entering the Olympiad just because they're girls, that is one thing to address, but if congressive people are being put off, then I'd rather address that by building and valuing more congressive math activities.

These sorts of activities crucially encourage depth rather than speed. Math contains deep ideas and uses logical processes, both of which operate slowly. Much of the point is the process, and if we focus ingressively on getting to the outcome quickly, then we will miss that point. There is a myth that math is about getting the right answer, but that's a very ingressive approach to it; a more congressive approach, and one that is much more prevalent in higher-level math, is that it's about how you construct rigorous arguments to show that something is an answer. A congressive approach says it's more important to learn that discipline than to learn the answer; it's also a more transferrable skill. Math detractors complain that they never have to use school math in their daily lives, which is probably true if the whole focus of their school math was on specific answers about things like triangles and quadratic equations. Unfortunately, math as presented in school is largely the ingressive aspects: getting the right answer, following rules that are imposed on you, a proliferation of facts that you're supposed to know, and solving problems. If instead the math were congressive—about building arguments, seeing relationships between things, and understanding contexts in which different things are true—then the math would be much more relevant to daily life and much more about genuine mathematical ability.

Some more ingressive people do just fine in the traditional math environments. I suspect that congressive people are put off school math, as it is presented so ingressively—both in the sense of presenting only the ingressive aspects of the subject and in the sense of presenting them ingressively. Furthermore, if boys are also pushed to be more ingressive than girls, then this would contribute to there being more men than women doing math in universities; conversely, art is more congressive in many obvious ways, and this might contribute to there being more women than men at art school. This is of course aside from the biases that are directly about gender, including discrimination, unequal expectations, stereotype threat, and the lack of role models.

I don't think we should make more congressive math environments just to pander to congressive people; rather, we should create those environments in order to nurture congressive behavior because of its value. I think it is valuable in life, and in particular in math and science at all levels. A broad example of this comes from Finland.

CONGRESSIVE EDUCATION IN FINLAND

The Finnish education system came into international prominence when a new large-scale assessment of secondary school students across the world was performed and Finland came out on top, surprising everyone including themselves, apparently. So global attention turned to Finland to see what it is they had been quietly doing.

Many surprises were in store. It turned out that the Finnish education system is essentially congressive. It focuses on collaboration, cooperation, and student well-being, on equity rather than excellence, and values teachers and their autonomy. It also features very little homework, shorter school days with many breaks, and vacations.

This is very different from the previously lauded East Asian model of drilling, academic pressure, and huge quantities of homework from primary school onward. It is described in *Finnish Lessons: What Can the World Learn from Educational Change in Finland?* by Pasi Sahlberg, director of the Finnish Ministry of Education's Center for International Mobility. Important features he discusses include an absence of standardized testing; a focus on cooperation, not competition; free school meals, health care, and counseling; not starting formal schooling until the age of seven; a generally relaxed and unregimented school day; the fact that there are no private schools in Finland (charging school fees is illegal).

A crucial part of all this is the system for selecting and monitoring teachers. Broadly speaking, in Finland teaching is a good job, with teachers there having "equal status with doctors and lawyers."* More to the point, teachers have professional autonomy, the curriculum is only laid out in broad brush strokes, and there is a lot of support for children who need extra help. Teaching is thus a profession that many

---

*LynNell Hancock, "Why Are Finland's Schools Successful?," *Smithsonian Magazine*, September 2011.

people actively want to enter. Potential teachers are chosen carefully, trained extensively (at the state's expense), and then allowed to have responsibility for their teaching instead of being saddled with accountability. Sahlberg says, "There's no word for accountability in Finnish. Accountability is something that is left when responsibility has been subtracted."

Previously, the received wisdom might have said that this is all very well, but the students wouldn't learn as much; this possibility is contradicted by Finland having come out on top in the first three iterations of the worldwide Programme for International Student Assessment (PISA), which focused on reading, mathematics, and science in turn.

PISA is a wide-ranging system for testing a sample of fifteen-year-olds. It doesn't just test their proficiency but also their motivation, attitudes, and anxiety levels. In terms of proficiency, Finland came out on top in the first round of literacy tests, of mathematics tests, and also of science tests. This "top" was rated according to which countries had the smallest percentage of students performing at the lowest of six levels, but even if you switch to looking at which ones had the largest percentage performing at the highest level, Finland was only marginally outranked in each case—in literacy and science by New Zealand, and in mathematics by Korea and Hong Kong–China. The United States was very far down each of the tables. Unlike Hong Kong, Finland also has one of the smallest differences between the strongest and weakest students (while still having one of the highest averages) and one of the lowest correlations between family background and eventual proficiency (unlike the United States).

There are plenty of people who are so conditioned into the ingressive style of education that they refuse to admit that the Finnish system could work anywhere else. They insist it must be some sort of anomaly, only possible because Finland is so small and homogenous, or because of some sort of Finnish genetics, or because the weather is so bad that nobody does anything except sit at home doing their homework.

But Finland is only as small and homogenous as many states in the United States, where education is run at the state level in any case. Moreover, Finland has had an influx of immigrants in recent years, meaning that some areas have become a lot less homogenous than others, but the differences in homogeneity don't seem to have much impact on the schools' outcomes. And they don't really have homework.

Incidentally, Sahlberg points out that they're not that interested in their PISA success, as they're interested in educating children to learn, not to take a test. They're more interested in equity than excellence, but it's funny how they focused on equity and achieved both greater equity *and* greater excellence, where other countries that focus on excellence achieve neither. Focusing on the good of society resulted in more highly proficient individuals as well.

There is a slightly sad sequel to the Finland story, which is that after becoming the focus of international attention, Finland's PISA scores started dropping, leading some people to gloat that it was all a mirage or a fluke. (Still, at twelfth place in math in 2015, Finland remained far ahead of the United States, who were in fortieth place.) Sahlberg gives various highly plausible reasons for this, including a sort of feedback loop—a system that did not previously aim for success in tests did well in some tests and then became self-conscious about it. Sahlberg says that the country lost the drive to keep improving the education system, whereas other countries began deliberately aiming for higher PISA scores. Nevertheless, Sahlberg maintains a congressive approach and says that the best thing to do is not adjust the system to aim for higher PISA scores but enhance equity and equality (which have also dropped in Finland's recent results). He says, "The Finnish way of thinking is that the best way to address insufficient educational performance is not to raise standards or increase instruction time (or homework) but make school a more interesting and enjoyable place for all."*

HOW THIS RELATES TO GENDER

One way in which Finland still sets itself apart from almost all others is in gender equality. I suspect that many problems around gender bias in other countries can be traced back to an ingressive education system designed, unconsciously and otherwise, to promote inequality.

Most countries have boys doing better than girls in math and science and girls doing better in reading, but Finland has girls doing better than boys in math and science as well, according to various measures. Digging into the specific global results yields interesting details, though.** Science proficiency

---

*Joe Heim, "Finland's Schools Were Once the Envy of the World. Now, They're Slipping," *Washington Post*, December 8, 2016.
**_PISA 2015 Results: Excellence and Equity in Education_, vol. I (Paris: OECD Publishing, 2016).

is divided into various categories, and gender differences are measured in each one. Boys do better than girls on average overall, but not in every category. Boys do better in content knowledge, but girls do better in "procedural and epistemic knowledge"; boys do better at "explaining phenomena scientifically," but girls do better at "evaluating and designing scientific enquiry" and slightly better at "interpreting data and evidence scientifically." It seems that girls are doing better at the more congressive aspects of science.

In fact, there are also data on gender differences in students' epistemic beliefs. The students were asked how much they believe in things like "The ideas in science sometimes change," "Sometimes scientists change their minds about what is true in science," and, rather crucially, "It is good to try experiments more than once to make sure of your findings." Girls were found to believe in all these statements slightly more than boys. Again, I think these are congressive aspects of science that are perhaps often neglected by the general public and general presentation of science.

The last statement in particular seems related to the idea of self-confidence. If you are overconfident, then you are less likely to check your answers and indeed your experiments. The PISA study even provides scope for investigating this, as along with assessing students' proficiency it included questions about students' attitudes. The results on "self-efficacy" are fascinating.* This is essentially a measure of self-belief, as students were asked questions about how easily they think they are able to do certain things. It will surprise few people

---

* *PISA 2015 Results*, I:137.

that boys turned out to have more confidence in their ability than girls, and it's tempting to conclude that self-confidence leads to higher achievement, but we must remember that correlation does not imply causation. In fact, it is to be expected that those who are better at math and science should believe that they are better and that those who are worse should believe that they are worse. The fascinating thing is that boys have more self-belief than girls even in Finland, where girls are actually better than boys. Moreover, although self-belief did mildly correlate with proficiency within countries, it did not do so across countries: some of the highest levels of self-belief were reported in the United States (who do not actually do very well in proficiency), and the lowest levels were reported in Hong Kong–China, where there were much higher levels of proficiency overall.

This was studied even more specifically by three researchers who used some questions specifically to test "bullshitting" in mathematics.* In a marvelously satisfying ruse, they asked students to rate their own knowledge of sixteen math concepts. The ruse was that three of the listed concepts were fictional: "proper number," "subjunctive scaling," and "declarative fraction" (the experiment was only done on students in English-speaking countries). Thus anyone who reckoned they were very knowledgeable about these three things was deemed to be "bullshitting." Again, it will surprise nobody that boys were found to bullshit more than

---

* J. Jerrim, P. Parker, and N. Shure, "Bullshitters: Who Are They and What Do We Know About Their Lives?," IZA Institute of Labor Economics Discussion Paper 12282, 2019.

girls, but the fascinating thing was that in North America the gender difference was much less—apparently American and Canadian girls have learned to bullshit just as much as their male peers, but in Europe only the boys have learned to bullshit as much as the Americans.

Of course, some aspects of these gender differences really do have to do with the different ways in which society treats men and women: if men are told they are brilliant a lot, then it's no surprise that they believe they are brilliant. If society and in particular men are constantly putting women down, then it's no surprise that women will not believe in themselves. But if we use ungendered terminology, we can separate out the question of what we actually want to encourage and nurture, which is particularly important when we're thinking about education, both at the level of broad educational systems and at the level of individual schools and classrooms. And then we can ask more careful questions about which characteristics are beneficial.

Self-confidence can benefit or harm students in different contexts because it can work in opposite ways. It can help students persevere when they're struggling, which is important, but it can also enable students to overestimate their abilities, with the result that they don't work hard or seek help. It can lead people to jump to conclusions and not check their reasoning—in normal life this means people do not fact-check, in scientific research this means people do not check their data or their methodology, and in mathematics this means people don't check their argument. This happens at all levels of math education, from children who do their work quickly without checking and so have many wrong answers to researchers who write down that things are "clearly" true but then nobody else can work out why that is the case—and sometimes it isn't.

How much self-confidence is good and how much is counterproductive? If we look at successful mathematicians and scientists, we find a variety of possibilities.

THE ROLE OF SELF-CONFIDENCE

The extraordinary and brilliant Indian mathematician Srinivasa Ramanujan came to wider attention outside mathematics following the film about his life, based on the book *The Man Who Knew Infinity* by Robert Kanigel. Ramanujan failed his (ingressive) high school exams and failed to get into university several times, but he persisted in pursuing research by himself and eventually became recognized as one of the greatest-ever mathematicians, albeit one with methods that were unconventional from the rigid point of view of European academia. His break came when Cambridge mathematician G. H. Hardy was able and willing to recognize the insight in Ramanujan's work in its own right, independently of external (ingressive) measures like grades and degrees.

In order for Ramanujan to persist in his research despite multiple external setbacks, he had to have a certain personality, a huge belief in himself, and the conviction that he was brilliant and that the world was just not recognizing him yet. I have met other (much less brilliant) mathematicians who have this conviction despite multiple failures at exams, grad school applications, and job applications, and they have all been men. In my experience of helping students, women are much more likely to conclude after one initial setback that they are not good enough. Of course, it doesn't help that they may have men all around them explicitly telling them they are not good enough (even when they are). And of course, it's not just women: it's congressive people.

The obvious conclusion to draw is that self-belief and self-confidence are valuable assets for a mathematician. But there is, as usual, something more nuanced going on. There are large quantities of people out there who really aren't brilliant at all, who fail according to all external measures because they aren't brilliant, but who still persist in believing that they're brilliant and that the world is just unfairly not recognizing them yet. Prominent physicists and mathematicians receive a huge quantity of unsolicited mail from these types of people, who are convinced they are a great genius who has solved everything despite the fact that they have no job, no research record, and possibly no degree or training in the subject. It is self-belief and resilience to failure taken to an extreme: ingressivity taken much too far. I have only ever received one such letter from a woman, and the rest have all been from men. Physicist John Baez calls them "crackpots" and has devised a "crackpot index" with which to assess how far gone they are. It is tongue-in-cheek but also has a large element of truth in it.

At the other end of the scale there are congressive students who are so aware of their own weaknesses that if they don't come in first on something, they might give up, believing that they're not good enough. They are the kind of students who don't imagine they're good enough to go to college, or don't imagine they're good enough to do a PhD, or don't imagine they're good enough to have a career as a mathematician. If I hadn't been accepted into my first choice of grad school, I would have given up—I had fully decided that I would go and do something else instead, because not getting accepted to my first choice would be a sign that I wasn't good enough. Note that this isn't just imposter syndrome, which is where someone continues to doubt themselves despite external measures of success, and fears they will be uncovered as a "fraud"; for women and minorities it might be because people have actively been doubting them all their lives, even when they have external markers of success to show, so their self-doubt is founded in external measures and is not just something inside their head.*

One possible conclusion is that we should get everyone to be more ingressive so that they don't give up. But then we'll end up with more people not giving up who *should* give up, and we also risk making congressive people feel uncomfortable or perhaps being put off by the very idea that they have to be more ingressive to succeed. Another possible argument is that we should get people to have a better balance between ingressivity and congressivity so that they don't give up so easily but also don't persist unrealistically.

But I am going to argue that there are important ways that congressivity benefits *individuals*, that they would lose that benefit if we managed to get them to be less congressive, and that the benefits of ingressivity are things we could replace with structural measures so that people don't lose out if

---

*Thanks to mathematician Dr. Marissa Kawehi Loving for drawing my attention to this exact subtlety.

we encourage them to be less ingressive. I would rather work with or teach someone who underestimates and doubts their own abilities rather than someone who overestimates them.

Like with the "dandelion" and "orchid" children, I would rather build congressive environments and structures to help congressive people flourish, instead of trying to get them all to be more ingressive. In the case of congressive students, one way in which we can help is to find the ones who underestimate their abilities and provide them with the validation they need in order to keep pursuing what they're trying to do. The only reason that their congressivity might be a hindrance is if the people around them do not provide that kind of support. Besides, some people might not want (or be able) to become more ingressive.

This is a way in which we can address gender imbalances directly, by specifically finding women who underestimate their own abilities and who may have been put off by peers or professors belittling them; I am grateful for the support I have received in this vein. But it doesn't just have to be about gender. I think congressive people have the potential to be better mathematicians, scientists, and general contributors to society if we nurture them well. In fact, many brilliant scientists talk about their own self-doubt.

CONGRESSIVITY MAKING SCIENTISTS BETTER

Dame Jocelyn Bell Burnell is a British astrophysicist whose doctoral work resulted in a Nobel Prize—for her supervisor (who happened to be male). She was not bitter about it, acknowledging that supervisors usually do get credit for the work graduate students do under their supervision, and that Nobel Prizes should not go to graduate students. She has proceeded to have a stellar career and in 2018 won a Special Breakthrough Prize in Fundamental Physics for "fundamental contributions to the discovery of pulsars, and a lifetime of inspiring leadership in the scientific community."

It seems that her "fear of flunking made her meticulous."* I read this as congressivity improving one's rigor as a scientist. Before I submit a paper for publication, the fear of a referee finding something wrong with it makes me check it even harder. Before I give a talk, the fear of being challenged by someone in the audience makes me prepare much better and anticipate all possible questions. Incidentally, the Special Breakthrough Prize comes with $3 million, significantly more than the Nobel, and Bell Burnell donated the entirety of it to help students who are female, underrepresented ethnic minorities, or refugees to become physics researchers, an extremely congressive act.**

Large prizes are a rather ingressive way to celebrate people, especially when the prize has to be awarded to individuals rather than groups. In his memoir *The Gene Machine*, Nobel Prize winner Venki Ramakrishnan points out how out-of-date the Nobel Prize is in its self-imposed limit of naming a maximum of three winners. That was not unreasonable at

---

* Sarah Kaplan and Antonia Noori Farzan, "She Made the Discovery, but a Man Got the Nobel. A Half-Century Later, She's Won a $3 Million Prize," *Washington Post*, September 8, 2018.

** Pallab Ghosh, "Bell Burnell: Physics Star Gives Away £2.3m Prize," BBC News, September 6, 2018.

a previous time in the history of scientific research, when collaboration was slow, unwieldy, or impossible because of the difficulties of communicating across long distances. But the ease and speed of modern global communication have enabled collaboration with no apparent limitation on distance or scale, and the three-person limit seems increasingly absurd. The 2016 detection of gravitational waves involved more than a thousand scientists across the world, but the Nobel Prize could name only three: Rainer Weiss, Kip Thorne, and Barry C. Barish. A 2015 paper on the Higgs boson lists an astonishing 5,154 authors; the paper consists of nine pages of research together with twenty-four pages of authors' names and institutions.[*]

Mathematics can still be done alone, although it is increasingly done in collaboration. Fields Medalist Terry Tao talks about how he was the ultimate ingressive mathematician when young (although he doesn't use that word). He was a classic "prodigy" who sailed through levels of education and excelled in competitions. But in his adult life he has become extremely collaborative and has written papers with an extraordinary number of other people—a hundred or so as of 2019. In one case he set up an open, online collaboration that made rapid progress on a problem as "people with different skills squeezed out what improvements they could."[**] The

---

[*] G. Aad et al. (ATLAS Collaboration, CMS Collaboration), "Combined Measurement of the Higgs Boson Mass in pp Collisions at $\sqrt{s}$=7 and 8 TeV with the ATLAS and CMS Experiments," *Physical Review Letters* 114, no. 19 (2015): 191803.

[**] Erica Klarreich, "Together and Alone, Closing the Prime Gap," *Quanta Magazine*, November 19, 2013.

activity was sparked by a result from a very different style of mathematician, Yitang Zhang, who was almost unheard of, and who more closely fits the old stereotype of a solitary genius toiling away obsessively by themselves. Zhang is quoted as saying, "I choose my own way, but it's only my way."* Tao, on the other hand, is somewhat more direct about discouraging young mathematicians from that route. Another congressive facet is that he is a prolific and generous blogger, and in one post he wrote about the solitary, obsessional path as a "particularly dangerous occupational hazard."** In another post he advises being very skeptical of your work, most of all when you're dealing with a difficult problem.***

This is sheer speculation, but perhaps Maryam Mirzakhani is another example of a mathematician who was fast and competitive when young and then became "slow" in order to be a deeply brilliant researcher as an adult. The skills needed for research are not the same as those needed for Olympiad-like competitions or even exams. Research involves collaboration, reference to all the existing literature, and doing things nobody knows how to do yet, in a timeframe, usually, of years. There are many examples of great researchers who were not good at exams. I think our focus on exams and competitions is missing the point and is sometimes actually harmful.

---

*Klarriech, "Together and Alone."

**Terry Tao, "Don't Prematurely Obsess on a Single 'Big Problem' or 'Big Theory,'" n.d., terrytao.wordpress.com/career-advice /dont-prematurely-obsess-on-a-single-big-problem-or-big-theory.

*** Terry Tao, "Be Sceptical of Your Own Work," n.d., terrytao.word press.com/career-advice/be-sceptical-of-your-own-work.

One example where competition in mathematics was obstructive was the long-running competition between Newton and Leibniz about calculus in the late seventeenth and early eighteenth centuries. They had different approaches and were fiercely competitive about whose approach came first. Arguably Newton's insistence on his supremacy, in combination with his extremely revered status in Britain, held back British mathematics for a hundred years, as British mathematicians were stuck using Newton's notation for differentiation where Leibniz's was much better.

What if we pooled our knowledge and collaborated instead? In small fields of mathematical research that is a much more common way to deal with different people working on the same thing in different parts of the world. This has become especially widespread now that communication is so easy and fast.

In my field of research, category theory, I have never seen a mathematician race to beat someone at their research. If someone is already working on a problem, we are more likely to leave it to them in order to avoid that type of competition, or to collaborate with them if we think we can help. Physicists and mathematicians have been sharing their work online for free more or less since the technology to do so first existed. We post articles online long before they are peer-reviewed, and this helps us get wide feedback, which in turn helps us improve our work. This is a congressive way of doing research. Keeping it secret until a big shock announcement is ingressive.

Many congressive mathematicians blog about their work almost in real time, sharing it as widely as possible while it is going on in order to get as much collaborative input from the rest of the world as possible. In fact, in one contemporary field related to category theory, homotopy type theory, an entire book has been collaboratively written using online technology, and it seems that the whole process, including dissemination, has been deeply congressive; one of the (more than forty) authors, Andrej Bauer, wrote on his blog, "The spirit of collaboration that pervaded our group . . . was truly amazing. We did not fragment. We talked, shared ideas, explained things to each other. . . . The result was a substantial increase in productivity. There is a lesson to be learned here."*

The general idea that sharing things makes them less is an ingressive mindset that leads to competition and individualism. I have been looking at math and science in both education and research, but this applies more broadly. Many other structures of society are set up as if they are driven by scarcity, when as with education there is no real scarcity. I am going to argue that these structures are based on assumptions that ingressivity is better, and that we thus think that filtering people ingressively will select the "best," when in fact congressivity might be much better.

I argue that ingressiveness is not actually better for society and so society would be better off not favoring it. If we try to fix gender issues without addressing this, then the fix will be superficial, like increasing diversity without improving inclusion.

---

*Andrej Bauer, "The HoTT Book," *Mathematics and Computation: A Blog About Mathematics for Computers*, June 20, 2013, math.andrej .com/2013/06/20/the-hott-book.

INGRESSIVE STRUCTURES

Society favors ingressive behavior structurally, in the form of systems that we use to run things and filters that we use to select people for those things. This can happen because of individuals who consciously favor ingressive behavior, but it can happen even when they don't. I have found myself accidentally defaulting to ingressive structures, and on one occasion it was actually when I had just given a talk about ingressive and congressive traits at a big conference.

There were several hundred people in the room, and at the end I announced that I had forty preview copies of my book to give away. There was a gratifying stampede of people trying to get to the front to get a free book, and then a gratifying moment of sheepishness as they all realized that they had been rather ingressive about getting a free book ahead of everyone else, while the more congressive people had stayed at the back. I started wishing I had come up with some sort of congressive method for handing out the books—perhaps an invitation to people to come and pick up a book to give to someone in the room who they felt really deserved it—but I had been rather congressive myself and not imagined that so many people would want the book.

Anyway, it brought into focus how often in life we reward ingressive behavior, even accidentally. And even if we do it deliberately, we might not have done it *consciously*. Ingressivity is so ingrained in our upbringing, in our education, and in society that we might just think we're rewarding something that's "good," such as with the "strong" arguments being favored in Oxbridge exams (as described in Chapter 3), without stopping to think about whether there's a congressive version of "good" that we're overlooking and that might even be better. Other times we might just never have thought about it, such as in first-come-first-served situations. For example, when people raise their hand to ask questions, why does the person who raises their hand first deserve to ask the first question? As mathematicians, we challenge the apparently basic principles and try to find more basic ones.

The fact that ingressive behavior is better for oneself and that congressive behavior is better for society is more or less built into the definition. But, as with most things, there is more nuance to this point. If you accept that society is made up of a collection of individuals, then the character of society broadly comes from two things: the characters of the individuals and also the structures they have for interaction. But individuals' behavior is in turn affected by society.

Perhaps there is a cycle of interactions, like this:

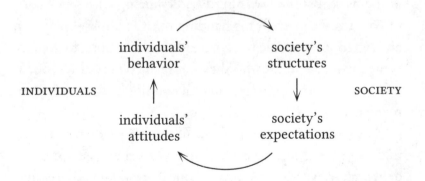

If we accept that society has an effect on individuals, then we see that what is good for society is also good for the individuals who are part of that society. In that sense, congressive behavior is good for society *and thus also* good for individuals in that society, whereas ingressive behavior can be good for individual people in society without being good for the whole. It is then a frustrating fact that although congressive behavior is better for society, our society is set up to reward ingressive behavior.

One ingressive filter that is very commonplace in society is competitions: explicit competitions with prizes, and competitive structures such as exams and elections. We use them even for congressive roles such as research, teaching, management, and playing music. Arguably, running a country also ought to be a congressive role, but we still select for it using ingressive systems; I'll come back to the ingressivity of politics in Chapter 7.

Through ubiquitous competitions we put pressure on people to be competitive if they want to succeed, or to give up on success if they don't want to be competitive. This starts in education, as we've discussed, but continues all across different aspects of society. It is no surprise, then, that people do tend to be competitive, but this does not show that such behavior is innate or "natural." I'm going to examine our assumptions around competition in a way that we could apply to any aspect of ingressive behavior. We will question whether or not competition is innate, what cycles are causing us to favor it, and what we stand to gain if we change our assumptions and our structures that perpetuate it.

In *No Contest*, Alfie Kohn conducts a comprehensive review of a vast quantity of research on competition. He debunks many myths, including the idea that competitive behavior is innate, that women are innately less competitive, and that competition is even good for anything at all.

As usual, one way to study innateness is to look at children's behavior. My nephew, when he was about five, asked in that earnest way children do, "Why did they have to have a world war? Why couldn't they just have a world meeting and sort it out like that?" And apparently it's not just him. Kohn tells of various ways in which children are observed to prefer collaborative games to competitive ones. If children are exposed to competitive games, they learn to become more and more competitive, but Kohn finds no studies showing a preference for competition once children have experienced cooperation. Preference is not the same as innateness, but if children prefer to be cooperative, this does throw doubt on the idea that people are "naturally" competitive.

Of course, children are exposed to competitive games all around them. In many cultures, particularly in the United States, they are encouraged into competitive sports in a way that they are not generally encouraged (in such a ubiquitous way) into collaborative music, theater, or other creative arts. One justification is that it's a way to pay your way through college (at least in the United States), but then the question remains: Why is it possible to pay your way through college by means of competitive sports but not music or theater?

Note that there are scholarships for musicians to study music, but the difference is that you can get an athletic scholarship to pay for your degree in *something else*. Why can't you get a music scholarship to pay for your math degree?

We could answer pragmatically that it's because sports generate far more money than music or theater, so universities make much more money from it and thus they can pay the student participants more and, most importantly, want to do so in order to attract the best athletes away from other universities.

Let us keep questioning this system to try to get further back into first principles. Why do sports bring in so much more money than music or theater? Because more of the general public enjoys watching them, and either pay for it directly or contribute to huge home viewing figures that then generate advertising revenue. Why do people enjoy watching sports? Well, that comes back in a big circle to them being exposed and pushed toward competitive sports as children.

Once we see we're caught in a cycle, we can ask ourselves whether it's a good cycle or not. As it's not based on anything, we can consider breaking it. In Chapter 7 we will come back to ways in which we can change structures to be more congressive; in this case it would mean building in structures to offset the pull of competitive sports as a way of paying for a university education. The most obvious congressive structure that addresses this is having more affordable or even free public education, which is probably why athletic scholarships are not such a skewing influence in Europe, where university education is often state-funded. We've come back in a circle to education again.

CONTRIVED SCARCITY

As we have described, Alfie Kohn characterizes competition as something that happens when resources are scarce. From here we get the idea that competition is a "natural" behavior, because "in nature" resources are scarce and so we have to compete for them. The thing is that, as with education, in many situations in modern life there isn't any real expend-able resource at play, but a scarcity is fabricated in order to create artificial competition. This seems to be not only be-cause we believe competition is good but also because we believe people simply enjoy it.

And so things are made into a competition by contriv-ing a scarcity. Children are put into music competitions in order to rank them even though playing music well is not a limited resource. They are put into math competitions, spell-ing competitions, general-knowledge competitions. It is said that this makes it "fun." But it only makes it fun for children who like competition, and plenty don't, including me.

A vivid example is the children's game musical chairs, in which you deliberately start with one fewer chair than chil-dren so that children have to compete for the chairs. At each stage you remove the "loser" and you also remove one chair so that you contrive scarcity for the next round as well.

In adult life this manifests itself in all the singing, danc-ing, baking, or "survival" TV shows in which one person is kicked off the show each week until there is only one winner. This is a form of contrived scarcity because "winning" is not a genuine scarcity—the only reason there can be only one winner is that someone has decided it to be so. The scarcity is contrived rather than real for most prizes, unless the point of the competition is, say, to select an architect for a particular building project.

In 2019 the Booker Prize was notoriously awarded jointly to Margaret Atwood and Bernardine Evaristo. This caused some general outrage because that was "against the rules" or because "failing to pick one winner shows you have failed."

Failed at what? Well, you've failed at picking one winner, but what was the point of picking only one winner except that the rules said so? Where did the rules come from? Charlotte Higgins wrote in the *Guardian*, "Everyone agrees that competition is the enemy of art. And yet, on the whole, there is also an agreement to conspire in the notion that it isn't."* She gives some examples of recent prize-winners who shared their prize with their fellow nominees: artist Helen Marten (the Turner Prize) and writer Olivia Laing (the James Tait Black Prize). In fact, in 2019 the four shortlisted artists for the Turner Prize requested to share it before it was even awarded, and the jury agreed. How does this relate to gender? Is it significant that two of the early winners to share their prize were women? Either way, I would say that sharing your prize is a very congressive act. Higgins ends her article by predicting that more artists and writers will take this congressive approach to prizes in the future, and suggesting that prizes themselves have the option of adapting to this congressive mood or losing their cultural dominance. She doesn't use the word "congressive," but I see it in this light. If I said it was a feminine thing to do, then that would do an injustice to any congressive men who might feel inclined to do similarly.

So much for the idea that competition is an inevitable product of society. What about the idea that competitive behavior is genuinely valuable? If we see that successful people

---

*Charlotte Higgins, "The Booker Prize Judges Have Exposed the Doublethink Behind Our Arts Awards," *The Guardian*, October 15, 2019.

tend to be competitive and that uncompetitive people tend
to be less successful, we might conclude that competitive
behavior is beneficial for success in practice. However, this
is one of those situations where there are additional root
causes contributing to that outcome of success, and if we
think harder about first principles, we can uncover them. In
this case, it's that society's structures are set up to reward
competitive behavior, so by definition competitive behavior
will yield greater success.

Is it fair to reward competitive behavior in this way? It is
instructive to challenge the idea that competition is beneficial. There are many ways in which competition could be actively harmful, and there are also many ways in which people
can be motivated other than competition, so whatever we believe is achieved by competition could actually be achieved in
some more congressive way. Finding those congressive ways
would have the double effect of avoiding the harm caused
by competition and also being inclusive toward congressive
people; in addition, we are going to see ways in which congressive approaches might be positively advantageous.

It might be tempting to point to many discoveries and
achievements that were driven by fierce competition, including in science, the space race, climbing Mount Everest,

reaching the South Pole, and so on. But it's important to balance things out in these discussions and weigh up the ways in which competition and collaboration might both contribute to and hinder progress, rather than simply pointing out things that have been achieved through competition. Could those things have been achieved without competition? As I mentioned earlier, my field of research, category theory, does not really involve people competing with each other for research. Many scientific discoveries came about through collaborative work rather than teams competing to be first, such as the recent first-ever image of a black hole. This accomplishment necessarily involved scientists and telescopes around the world so that as the black hole disappeared over the horizon for one telescope, others could start picking it up, and the images were then "stitched" together by supercomputer. The network of telescopes is called the Event Horizon Telescope and was founded by astronomer Shep Doeleman. More telescopes have joined in since that initial picture was created.

So competition is not always necessary for scientific achievement. Moreover, competition can have bad side effects. One meta-analysis found that almost 2 percent of scientists admit to having "fabricated, falsified or modified data at least once" and almost 34 percent admitted "other questionable research practices."* We should wonder whether they are driven by the competition to publish, which is in turn driven by the ingressive way scientists and institutions are evaluated on the basis of how many papers they have published.

---

*D. Fanelli, "How Many Scientists Fabricate and Falsify Research? A Systematic Review and Meta-Analysis of Survey Data," *PLoS One* 4, no. 5 (2009): e5738.

When universities compete in a fee-driven market, this arguably results in more expensive facilities and grade inflation, not necessarily better education itself (unless you think facilities and grades equate with quality of education). The competition among students to win places might have led to some students working harder, but it has also led to some parents employing more and more nefarious means to get their children a "prestigious" university place.

There is also a question of whether competition really spurs individuals to do better. Highly competitive situations can lead people to cheat, deliberately sabotage others, or give up. They might give up because they can never win (like me and sports), but they might also give up after the competition has ended because they did not develop any intrinsic motivation for the activity, only the extrinsic motivation for winning.

I am an example of someone who does not do better in competition but rather in collaboration. I am sure that there are many doctors and nurses who are motivated by wanting to help sick people, not just by competing for prizes or income. I am sure there are firefighters who are motivated by saving lives, not by trying to compete to save the most lives. What about teachers who are motivated by wanting to educate young people, and musicians who keep working to do better and better even when they're not in competition with anyone? There are men and women working in all these fields, but research shows that men tend to be more competitive than women. For example, one study in the *Quarterly Journal of Economics* found that 73 percent of male participants chose a competitive incentive scheme compared with only 35 percent of women.* The idea that men are more competitive than women is sometimes used to justify the dominance of men in certain industries or in positions of power, but this is a bad argument.

COMPETITION AND GENDER

In the framework of weak arguments I introduced in Chapter 2 there is a weak argument that looks like this if we keep thinking along gendered lines:

1. Men are observed to be more competitive than women.
2. Being competitive is better for success.
3. Therefore, men are more successful than women.

With this argument, we are in danger of being pushed into this dichotomy, as in Chapter 3:

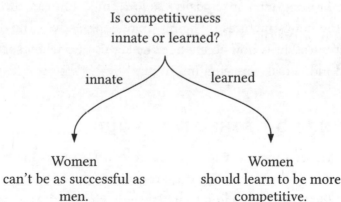

Is competitiveness
innate or learned?

innate learned

Women
can't be as successful as
men.

Women
should learn to be more
competitive.

---

*M. Niederle and L. Vesterlund, "Do Women Shy Away from Competition? Do Men Compete Too Much?," *Quarterly Journal of Economics* 122, no. 3 (2007): 1067–1101.

In this argument "competitiveness" could be replaced more generally with "ingressivity." Once we acknowledge that ingressivity is largely better for success *because society is ingressive*, then we can think more clearly through this argument instead of being pushed into the innate/learned dichotomy. Society favors ingressive behavior, so ingressive behavior seems necessary for success. Men are currently more ingressive and so they are more successful. Currently women are more congressive and so they're being excluded. Moreover, congressive behavior is better for us, so the exclusion of congressive people is bad for all of us.

Favoring women is one solution but keeps us stuck in the gender dimension. Favoring congressive behavior because it's better for us frees us from that trap. Incidentally, I also think favoring congressive behavior is actively feminist, because it's fighting against a status quo that favors men, and moreover it has the potential to be less divisive, because we're not fighting men specifically. Indeed, men too can succeed in the new structures by being more congressive. The difficult question is how to change the status quo when society, and individuals, are clinging to it.

CLINGING TO THE STATUS QUO

We've seen various ways in which, contrary to some widespread assumptions, being ingressive might not be beneficial. So why do we cling to it? If we do cling to it then we're helping to maintain the power structures of the status quo. We might say that if we're not the people in power, then we can't try to change the system, and that if the people in power don't want to change it, then it will never change.

However, if we believe we can't change anything, we really won't be able to. We should make sure we're not just telling ourselves we are powerless to justify our inertia.

General inertia holds us back in many ways. It is often much easier to carry on doing the same thing rather than try to change it, especially when you can't be sure what will happen if you do change it. You might not believe it's possible to change it, or you might not believe that the result will be better, like women who benefit from white men's power and so don't feel motivated to overthrow it. Then it seems to come down to the people who are most disadvantaged in the current system to do all the work of changing it, as they are the ones with the least to lose. But they are also the ones with the least power to change anything.

An added conundrum here is that it might seem that it would take ingressivity to dare to change anything, but if it's only the congressive people who want to change things, then we're stuck. The ingressive people who benefit directly from the ingressive system will of course always seek to keep it the way it is. I believe this currently largely benefits men, along with women who successfully emulate ingressive behavior. But again, it's not all men and it's not all women, so thinking in terms of ingressive and congressive behavior can help us become clearer about what is really going on and how to change it without just excluding men in return.

I am going to argue that change is possible and that almost everyone will be better off if we effect that change—in

fact, everyone except those who currently hold disproportionate power. This is about sharing power according to
character traits rather than fixed identities, and so I believe it
will empower women as well as any other underrepresented
and marginalized groups.

Crucially, I believe that we can make this change congressively. As a first step, we can start to break the cycle by
breaking the link between the ingressiveness of society and
our own ingressive behavior. We can find ways to operate
congressively even within ingressive society. This is the subject of Chapter 6.

CHAPTER 6
LEANING OUT

I EVENTUALLY REALIZED that I was unhappy with the standard ingressive academic environment, unhappy with how my character was being influenced by it, and unhappy with the effect I perceived it to be having on students. So I left the standard academic career path and built myself a congressive new career based on sharing my love of mathematics as widely as possible, bringing more people into math, and removing the obstacles and barriers around the world of math. I have left behind the ingressive principles of exclusivity, evaluation, competition, and "brilliance" that I didn't like in mainstream academia. In my congressive new career it so happens that I work with a vastly higher proportion of women than I used to. I now do far more work than before on outreach and inclusivity in math, and there seem to be far more women working in these areas than there are generally in research departments. Of course, they're not all women—they're largely congressive people, so I get to meet congressive people of all genders who care about these principles. Many of them are in fact professors in research departments, trying individually to fight against the ingressive environment on behalf of their students and the wider population.

Perhaps if I went back into the ingressive academic world now, equipped with my understanding of ingressive and congressive behavior, I could work out how to deal with that world without becoming as ingressive as the environment. Better still, perhaps now I could work out how to change the environment so that it would become more congressive, and thus encourage everyone in it to become more congressive as well. As Ruth Whippman writes, "Enough leaning in. Let's tell men to lean out."* But, again, it's not just men: sometimes it's women who are too ingressive, especially in careers like academic disciplines, where they may have felt the need to emulate such behavior to succeed and, having done so, might be rather proud of it and suspicious of any suggestion that another way is possible. But I would like everyone, men and otherwise, to become more congressive, and I believe that we can do this even inside our existing ingressive power structures, while we are still in the process of trying to change them.

At the heart of our existing power structures is a system of assumptions that is propping them up and keeping the balance of power away from congressive people. As I've said before, there is also a system of direct prejudice keeping women and minorities out of power, but I think it's beneficial to think about these issues along separate dimensions to make sure we address them both.

---

*Ruth Whippman, "Enough Leaning In. Let's Tell Men to Lean Out," *New York Times*, October 10, 2019.

The assumptions I'm talking about are that certain behaviors are important for success and that the best way to achieve those is something ingressive. This could be self-confidence, risk-taking, or resilience, for example. This diagram depicts that system of assumptions, with Y representing any of those qualities in question:

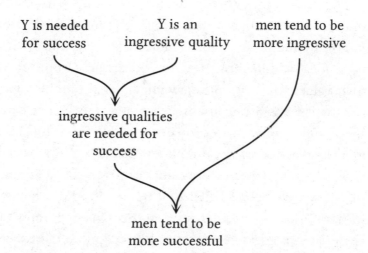

The three basic factors that seem to be causing the outcome are listed at the top. The right-hand side is gendered thinking, and within this way of thinking, women who wish to be successful must become more ingressive—essentially the idea behind "leaning in." However, if we consider the middle and left-hand factors instead, then we are moving away from directly thinking about gender differences. We might still get stuck in ingressive assumptions, but we have a chance to move away from that too. I am going to challenge the middle factor and argue that we can find congressive approaches to things that seem ingressive, such as self-confidence, and that maybe the congressive version is so different that it needs a different name. I will also address the left-hand factor and the possibility that actually some very different, more obviously congressive qualities can be better for success even while we are in a largely ingressive world. This is how I have managed to unlearn my learned ingressive behavior and build a congressive career that is much more in line with my true values. I meet many women who share those values and are put off by the pressure to be more ingressive, and I believe these ideas will help them participate without going against their values. But with our new ungendered terminology we see that this will not just help women. I have decided to speak up for congressive people who do not want to emulate ingressive behavior, as I think they are losing out in society.

With many types of opposite characteristics it is easiest to say that we need a balance of things. Would it be better to have a balance of ingressive and congressive traits, and encourage congressive people to become more ingressive as well as ingressive people to become more congressive?

I think it's unnecessarily divisive to ask which is "better"; rather, it is more productive simply to look at different situations and consider in what way each type of behavior is beneficial. I see only a few situations in which ingressive behavior is really important, and find that in most of those cases such behavior has downsides that go with it. Furthermore, even then the benefit could be achieved in a congressive way without the attendant disadvantages.

Anyway if, like me, you don't like being ingressive or feel you can't be, then the question of which is "better" is moot. It would be better if I were taller—but not too tall. Being short is largely inconvenient, except on rare occasions such as folding oneself into a plane seat or taking a nap on a sofa. However, it's a bit pointless saying it would be better to be taller, because I can't be taller. It's more productive to find ways to mitigate being short, like having a library-style rolling stepstool in my kitchen so that I can reach the higher shelves.

So if you feel that you can't or don't want to be any more ingressive, then it's more productive for us to think of congressive ways to deal with the world. Chapter 5 was about how the structures of society tend to be ingressive and how that nurtures ingressivity in people, which in turn makes society more ingressive. Now we're going to look at how individuals can be congressive even in an ingressive society, and how small-scale congressive interactions could gradually influence society to become more congressive as a result.

The study of how individuals interact with structures is widespread in math, although in that case the "individuals" are mathematical objects rather than people. We study large structures broadly, and we also zoom in and study the small pieces that fit together to make those large structures. We study different ways that the small pieces could fit together to make different large structures, and we also study ways that different small pieces could be used to build the same large structures.

One example of this is with maps. We can make flat maps of small regions of the earth such as an individual town, but if we stick them all together we need to add in some curvature in order to get a whole globe. If we stick them together too simply, then we will make a flat earth, which—despite the beliefs of a few deluded people—is not correct. However, it's useful to have small flat maps of local areas because most of the time we humans are only moving around small areas of the world.

The interaction between local properties and global properties is a big part of math because we often try to build up our understanding of situations by starting with small building blocks and then building them up gradually into larger things. Then, faced with a large and complicated structure, we can try to break it down into small parts and understand how they fit together into the whole.

One of the powerful aspects of category theory is that it works very well as a framework at zoomed-out levels and also zoomed-in levels. We can look at relationships between whole worlds, relationships between objects within worlds, relationships between different parts of an object, and so on. This flexibility of levels is an advantage of abstract theories. In this chapter I'm going to zoom in and look at how we as individuals can operate congressively even within ingressive society.

I offer this as an alternative to the idea of "leaning in" for more congressive people who aren't comfortable with that idea. We have previously seen that when we are still stuck in the dimension of gender we get caught in a trap of thinking that women either have to become more like men and "lean in" to be successful or resign themselves to not being successful. Or we try to address male domination by trying to get men to be "more like women." Now we can rephrase the problem like this: Everyone might think they have to learn to be ingressive or else give up on success. Some people are more comfortable learning ingressive behavior than others. I did it for a while and then realized I was not comfortable with it, but this did not mean resigning myself to not being successful. There is a third way, which involves neutralizing the ingressive pressures of society and finding success congressively.

First I would like to present some role models, people who have found ways to become more congressive over the course of their life, to help convince us that doing so is possible and worthwhile.

CONGRESSIVE ROLE MODELS

I found Michelle Obama's memoir *Becoming* eye-opening in many ways, but I'd like to focus on the part of her story that seems to me to be about ingressive and congressive versions of success. She writes of pursuing success first according to the usual definitions of success: going to a prestigious law school, getting a job at a prestigious law firm, earning a huge salary. She achieved those things, but then she realized that wasn't what *she* wanted for herself. Perhaps she felt she had to do it first to prove that she, a black woman from modest beginnings on the South Side of Chicago, could do it. But then she wanted a different kind of "success": success evaluated on her own terms in work that she found meaningful. She asked herself, "How do I want to contribute to the world?" and realized that the job she had as a hot-shot lawyer wasn't it.

She spent some time investigating ways she could make a more meaningful contribution with her expertise and

qualifications, and then she switched careers, first working as an assistant to the mayor (taking a 50 percent pay
cut), then working at Public Allies, an organization recruiting promising young people who might otherwise be overlooked and giving them training and support for nonprofit
or public sector work. After that she worked on community relations for the University of Chicago to bring down
the walls between the university and the South Side neigh-
borhood that is all around it but so often excluded from it.
She calls these more "civic-minded" jobs rather than the
"*my-isn't-that-impressive* path."

I would say that she first pursued a traditionally ingressive career as a lawyer, was successful at it, and then realized
that she really wanted a more congressive career and a more
congressive form of success. She congressively went about
working out how to do that, thinking about her values and
consulting people she respected. It is telling that she first felt
she had to prove herself in the traditionally ingressive career
path. At risk of sounding like I'm trying to compare myself
with Michelle Obama, I will say that I too felt I had to prove
myself in the traditionally ingressive career path of academia
before I felt I could reexamine my values and create a more
congressive career for myself. Oddly enough, perhaps if I
had been more ingressive I would have felt able to reject the
traditional framework earlier.

In *The Art of War for Women*, Chin-Ning Chu says, "We
have trained our minds to think of success in a certain
way—the *male* way; it's only about getting ahead, climbing
the corporate ladder, becoming CEO." I would, of course,
call it the *ingressive* way, to extract ourselves from gendered

arguments. Extracting ourselves in this way can also help us see that we don't necessarily need female role models. We do need female role models to help us see beyond gendered oppression, but to help us see beyond ingressive assumptions we need congressive role models of any gender.

When looking for such role models we can run into a vicious cycle: if society celebrates ingressivity, then we will hear about ingressive people more. But other stories are there if we look for them.

Professor John Baez is a physicist and prolific blogger who started sharing his understanding of physics way back in 1993, when the internet was barely known. He wrote a weekly column, "This Week's Finds in Mathematical Physics," which was a congressive way of helping the world and himself at the same time: he knew that the best way for him to understand things for himself was to write about them for other people. Many great teachers find this about teaching— that you understand something much more deeply when you're motivated by the desire to help others understand it. This is one way in which congressivity helps oneself. It is very different from the kind of teaching that consists of posturing at the front of a lecture hall and relishing the power you have over the hapless students. Baez is so focused on helping society that he thought about what he could best do with his mathematical expertise to help the world, and he decided this meant shifting into working to save us from the global ecological crisis. John inspired me to think similarly deeply myself about how I could best use my mathematical expertise to help the world. Although my answer is quite different in specifics, it is the same idea.

When I began my career my ambition was to be a professor of mathematics at a top university, leading a powerful research group attracting brilliant PhD students. Now my ambition is to change the world to bring more people *into* things, people who have previously been excluded: from mathematics, from science, from education, from full participation in society. Remember, it doesn't have to be a competition, but I might actually say my new ambition is *more* ambitious than my old one. But I think it is a congressive ambition where the old one was ingressive. In making these changes to my career and life I was greatly inspired by John Baez.

I have needed role models who are women, to help me believe that it is possible to be a successful woman without emulating male behavior, but it is helpful for me to see congressive role models regardless of their gender, such as Baez. It's particularly important to separate out those issues because so many successful women who are held up as role models are in fact rather ingressive. Congressive female role models would be most helpful, but congressive male role models are much more helpful to congressive people than ingressive female ones.

One exception to this is the possibility of ingressive women who have consciously worked hard to become more congressive, for the good of people around them. Professor Emily Riehl is an abstract mathematician currently working at Johns Hopkins University. She also happens to have competed in Australian football at the international level, representing the United States, and by her own admission is extremely competitive and ingressive. This stood her in good stead when she was applying for grad school, making her way as a young researcher, getting her research known, and introducing herself to senior professors at conferences.

But as she became more senior she realized that a more congressive approach was important when she was teaching and nurturing new students. She runs many seminars and workshops for grad students and events for women and minorities, and stresses at the beginning that if she is being too ingressive about her teaching, then she really wants people to ask questions and get her to be more congressive. I recently asked her at a conference how she manages to ask so many questions during people's talks, which I still don't dare do in front of an audience in case my questions are considered stupid. It turns out that it might seem ingressive of her, but she's really doing it congressively—she knows that it helps other people and it also helps to change the tone of a talk, so she has committed herself to doing it in every single talk she goes to. She admitted that she was terrified of doing it at first but forced herself to do it anyway. One might call that an ingressive approach to being congressive—forcing oneself to do something unpalatable to help others.

It might seem easier to dare to change if you are more obviously ingressive, but personally I find that helping others is a very strong congressive motivation for doing things that seem daring. In fact, this is a specific example of a general principle: we can look for congressive ways of doing things that seem on the face of it to be ingressive, like daring to change.

This might involve reframing many things we take for granted, in further acts of digging down to first principles.

I have spent the last few years doing this in my life. I have found that becoming more congressive has not just been a substitute way to motivate myself to do ingressive-looking things; it has opened up far greater potential in me and has enabled me to make better contributions to society than when I was trying to be ingressive. This is aside from it simply being abstractly eye-opening to see the extent to which we make unnecessary ingressive assumptions about life. Also I am happier.

In the two-dimensional plane of ingressivity and congressivity I think I've broadly moved over the course of my life somewhat like this:

The gradual rise shows that as I learned to be more ingressive (competitive, combative, ambitious) I also learned to be more congressive (collaborative, nurturing, actively helping people and bringing people together). The peak shows the height of my attempt to respond to society's rewards trying to be "successful" in a standard academic career. Then when I started unlearning the ingressive behavior I didn't yet know how to be actively more congressive, and I'm still learning that.

Some people might be equally ingressive and congressive by being neither very actively, so they'd be very near the bottom left. Others might be very actively both things in different situations (or perhaps at the same time), so they'd be up at the top right. It's important to remember that in this two-dimensional version of a spectrum there are many different ways to be "equally" both things—anywhere along the dotted diagonal.

A full representation of the gradually shifting nuances of the two-dimensional spectrum is only really possible if we use two colors; this is the red and blue square incorporated in my author photo.* The two colors can gradually get darker as we move up or right, and they can also gradually blend together toward the middle.

Because there's not a clean dichotomy between ingressive and congressive behaviors, the shift to more congressive behavior can happen gradually by tiny increments. It doesn't have to be a sudden flip. It can also happen in some situations but not others, just as I became more congressive in non-work situations before I was able to become more congressive in work situations.

It might sound impossible to be congressive in an ingressive world without changing the structures first, and it might also sound impossible to change the structures, in which case are we stuck? What I do know is that if we assume it's impossible to change anything, then it will be impossible. I also know that small changes are possible and that they can

---

*If you're not reading a physical book with my author photo on the jacket, you can view it at eugeniacheng.com/square.

build up into big ones. So I'll start with small changes we can make to our own personal ways of thinking.

REFRAMING INGRESSIVE ASSUMPTIONS

As we're very used to our ingressive society, we may well be thinking ingressively without realizing it. It may seem that we have to be ingressive to succeed, but instead we can reframe some ingressive-sounding behavior so that it becomes congressive. We may have been looking at things from one direction only, given that through most of history all successful people have been men. This gives us a one-dimensional view of what success looks like, and has resulted in women emulating men's behavior in order to try to achieve equality with men.

In fact, it might take even more ingressivity for women to emulate the behavior of men, because men do actually experience the world differently. It might not take much daring for a white man to speak up in a group setting because he suffers no social threat and may feel no negative effects if he turns out to be wrong. By contrast, it might take much more confidence for a woman to speak up in the same situation if she is in genuine danger of being ridiculed, as women often are. She may also feel hurt by ridicule more than the man does, so even if people do laugh at the man for being "stupid," he might find it easier to brush it off. All this is to say that if women emulate men's behavior, we are not necessarily having the same experiences in the process.

Instead of ingressively facing such "risk" head-on, we can do something congressive about it. In the end I think it would be much better if everyone learned not to mock other people, but in the meantime we can think of ways to get through such situations congressively rather than just plowing through. I am aware that there are many ways in which I might look ingressive, but my motivations are actually congressive. I'm really not inherently self-confident, but I stand up and speak in front of large audiences. I avoid taking risks, but I took the apparently very risky step of quitting a secure, permanent job to become largely freelance. I am easily hurt, and yet I put myself on social media at the mercy of all the random obnoxious men on there (and yes, the people who are obnoxious to me on social media are almost always men). This is because I have found congressive ways to do things that look ingressive.

Understanding this possibility of reframing can especially help people who feel put off by ingressive situations, as happens with many women. It can also help us change the atmosphere in any situation to help congressive people participate. I think this is an important aspect of general inclusivity that will in particular help gender imbalance.

Take the example of "strength." It might initially seem that strength is more ingressive than congressive, but there are different ways to be strong. Traditional strength might be thought of as power, toughness, the ability to be unyielding either physically or emotionally, and thus to overcome others. That is ingressive strength.

Congressive strength could be more like flexibility. Some physical materials are strong precisely because they are flexible, whereas something rigid is more likely to break. Congressive strength is like a river altering its course to wind its way through a contoured landscape, as opposed to a water-fall that just pours out. Or deep-rooted plants that can sway in strong winds, compared with a big tree that is more likely to fall down.

Another congressive form of strength is to know when and how to get help, and to draw support from people around you. You don't have to be independent and self-reliant in order to be successful. Our ingressive society celebrates independence and "self-made millionaires." But nobody is really self-made. This is one way in which our gendered society can be detrimental to men, as getting help is seen as a sign of weakness in men. And thus many men, because of society's misguided pressures, miss out on the benefits of support, and even medical care.

Nobody accomplishes anything in isolation from the people around them and society at large. Building a good network of support is a congressive alternative to self-reliance, a way of building interdependence rather than independence. (It's also how I deal with the stress of obnoxious people on social media.) It can also be more efficient and even fun, like taking turns to cook for a group of people rather than everyone cooking for themselves alone.

Related to self-reliance are self-esteem and self-confidence. Those might sound like ingressive traits, but I think it depends where the self-belief comes from. Let us go a bit further into first principles and think about what self-esteem means. It can sound like it is belief in yourself that comes from nowhere but yourself, a bit like being self-made or self-reliant. That would indeed be ingressive, and

I've realized that those ingressive undertones put me off: I feel like I don't want that sort of self-esteem.

But belief in yourself can also come from external validation or preparation. If you achieve some things in the world, then it doesn't take self-reliance to start believing in yourself. Likewise if you have prepared very thoroughly for something. If you have achieved nothing or are woefully unprepared but still for some reason believe in yourself, then I would call that ingressive. Those are two different types of self-esteem. Here is a diagram depicting these different ways of having or lacking self-confidence, relative to external validation or preparation:

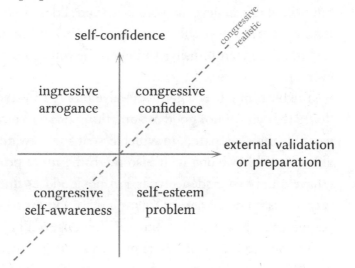

I definitely feel that the confidence I have comes from external validation and preparation. I dreamed of being a math professor but wasn't sure if I could do it. So I worked hard, and every time I overcame another hurdle I became more confident. But I still came up with long lists of backup plans in case I didn't make it. My confidence grew as more senior professors encouraged me, and I feel I gradually inched my way into the top right. By contrast, ingressive aspiring mathematicians keep believing they can do it even when rejected repeatedly; they are in the top left. Needing external validation isn't necessarily a sign of insecurity; it could be about making sure you're realistic rather than arrogant or delusional.

On the bottom half of the diagram are different ways of lacking self-confidence. If you're on the bottom right, you have plenty of external validation and are very well prepared but still don't believe in yourself—and I'd say you might have a self-esteem problem (or stage fright). Being congressive doesn't mean refusing to believe in yourself even when everyone else does.

The bottom left is where you have plenty of external evidence that you're not good at something or you know you're woefully underprepared, in which case it's self-aware and realistic to acknowledge it. It also might be more productive. I have plenty of evidence that I'm not good at finding my way by instinct, so I always follow a map; I know some other people who always get lost but *still* don't use a map.

The dotted diagonal line shows a sort of "balanced" congressivity: a view of oneself that is carefully rooted in external validation. Someone might argue that it's productive to be a *little* more ingressive than that so that you have some intrinsic sense of your own worth regardless of the unpredictable nature of the world's outcomes. That might be true, but then I'd also argue that I'd rather this were instilled in childhood by the right sort of validation from parents and educators, in which case that apparently intrinsic self-worth would actually be based in external validation as well. And understanding that things like exam results and prizes can be a bit unpredictable (and contentious) is, I think, part of having a balanced response to them.

In yet further nuance, there are ingressive and congressive forms of validation too. Ingressive validation takes the form of awards, prizes, grades, income, status symbols, and other external signs of your "superiority" to others. Congressive validation is more about the knowledge that you have made some sort of contribution, helped some people, created something, improved something.

A particularly pernicious form of ingressive validation is if you bolster your feeling of superiority by putting other people down. This can take the form of trying to make them feel inferior because of their knowledge, tastes, job, or social standing, or excluding them from something so that you feel special. Prejudice and bigotry are ingressive ways to bolster your self-esteem by keeping other people out of something. They involve declaring some other group inferior in order to declare your own group to be superior. In *Testosterone Rex* Cordelia Fine writes of a study finding that men's self-esteem went up when they were told they had done badly at something at which women tend to excel, because "incompetence in low-status femininity helps establish high-status manliness."

Arguably that type of need for feelings of superiority stems from deep insecurity—the opposite of self-esteem. Perhaps in that case it actually takes more self-esteem to be congressive and not need that feeling of superiority. And, of course, instead of associating character traits with gender and thus considering male ones to be superior, we can now look at them without associating them with gender and evaluate them for what they are.

Self-confidence that comes from external validation could be thought of as a congressive form of self-confidence, but the very concept of "self-confidence" sounds ingressive and might put congressive people off; it definitely put me off in the past. There are many other character traits widely considered important to success that similarly sound ingressive, but we can look for congressive versions of all of them. Another one that I've mentioned with respect to me quitting my job is risk-taking.

RISK-TAKING

Our ingressive society reveres and rewards risk-taking. At a superficial level we marvel at people who do things like climb up rock faces with just their bare hands or walk a tightrope between skyscrapers. At a less superficial level, high-risk research in science is often seen as more exciting. This might take the form of implausible experiments that show something quite unbelievable; it's not that there's a risk to anyone's safety, there's just a high risk of failure. We'll come back to ways in which low-risk research can be more valuable, and how to nurture those, in Chapter 7. Risk and gender inequality are linked in subtle ways, and a great deal has been written on that elsewhere (see *Testosterone Rex*, for example). It has been shown that women exhibit less risk-taking behavior in general but that this is not an innate biological difference. The aspect I want to address is the way in which women are then exhorted to learn to take more risks in order to be as successful as men; it's yet another form of "leaning in."

I'd like to give a sort of blueprint for how we can learn to "lean out" instead—that is, reframe apparently ingressive behaviors in a congressive way, and then take congressive action instead. This involves going back to first principles (again) and asking ourselves what it really means to "take risk."

Here is a diagram depicting some different responses to risk:

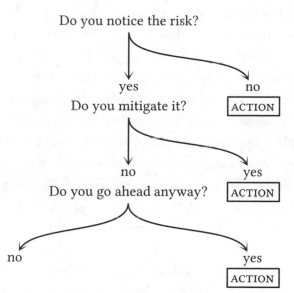

The left-hand side is where the risk actually prevents you from taking action, whereas any of the outcomes on the right-hand side (marked ACTION) could be described as "taking a risk." However, I think they have very different flavors. In the middle version you mitigate the risk, perhaps by building a huge safety net, in which case you're not really taking a risk at all, as all the risk has gone. I don't think this version is what is generally meant by "risk-taking," but in particular it's not what I *feel* when ingressive people insist that you must take risks to be successful; I think congressive people might feel they're being asked to do the top or bottom versions, where you either block out the risk or charge into it.

It's also important to remember that the same situation can involve genuinely different amounts of risk for different people, as we already discussed in the context of people speaking up in group situations, where there is genuinely more risk to women than to men. Big strong men really are at less risk than I am walking alone on the street at night. More abstractly, people with high social status can take bigger social risks, or, rather, those aren't actually risks at all.

Of course, ideally we would remove the risk by stopping people from doing that belittling or otherwise obstructively ingressive behavior. But in the meantime, what can congressive people do? If congressive people feel that they must take risks to be successful, they might be put off from even trying. Instead, I urge congressive people to build big safety nets to reduce the risk. This can be in the form of contingency plans, careful preparation, and a support network. And this doesn't just have to be defensive behavior; it can be actively productive too, and thus something that more ingressive people could benefit from learning. I am an extremely risk-averse person, but rather than hampering me, this has caused me to become much better at things through my efforts at reducing those risks. I honestly feel that I have not taken substantial risks in my life.

I invite you to question any received wisdom or advice about how to be successful and ask whether it is making ingressive assumptions. If so, and if you or anyone else finds that off-putting, you could try reframing the advice in a congressive way or finding a congressive way to approach the same sort of outcome.

An analysis along these lines can help us find congressive approaches to other ingressive-sounding attitudes or behaviors that ingressive society tells us we should learn, such as "leaving your comfort zone" or "resilience." Instead of leaving my comfort zone (which sounds terrifying), I work out how to extend my comfort zone. Resilience is also an off-putting idea to me because I don't like the idea of bouncing back or of being somehow impervious to bad things happening. I would rather be a sensitive human who is hurt by bad things but then tries to find ways to transform the bad experiences into something good, such as a way I can help other people.

Some people insist that that is a form of resilience, but if the whole notion of resilience puts some people off, then I think we should extract the part of the concept that isn't off-putting and call it something else, much as we did with "congressive risk-taking," which was really about building safety nets.

First, as with risk-taking, I want to stress that resilience or the appearance of it is much easier for privileged people. You can always weather misfortune better if you have more money (all other things being equal), or more social status (by being white, or male, or straight, or cisgender), or better health. We should be careful about whether asking people to become more resilient in order to be successful is insensitive to their disadvantages in life.

But I also worry that insisting on resilience can be used as a form of oppression, masquerading as a way to help marginalized people progress. Women are told that they need to be more resilient in the workplace, but this often means that women should just put up with bad behavior, bigotry, sexist jokes, and harassment. If I had been more resilient, I would have stayed much longer in a job I didn't like, thinking I should just put up with feeling undervalued, and bullied in ways that might have been racist, sexist, or ageist.

Being congressive in an ingressive environment does not mean putting up with sexism instead of fighting back; it's about learning to thrive, if possible, or figuring out how to get out if it's not possible to thrive. Perhaps "congressive resilience" is really about transformation. Rather than put up with an ingressive situation, we can try to transform it.

CONGRESSIVE RESPONSES TO INGRESSIVE ENERGY

How can we deal with an ingressive situation without either being overwhelmed by it or becoming ingressive in turn? How can congressive people make themselves heard in the face of incoming ingressive energy? I previously trained myself to be ingressive right back, but every time I did it I would go away and really dislike myself. And it didn't really achieve anything except perpetuate and escalate the ingressive energy in the world.

But that seemed like the only option other than just letting the ingressive people walk all over me, which might not have escalated the ingressivity but certainly enabled it. Congressive people often get walked on because we're trying so hard not to be a horrible person. In particular, women are pressured by society to sacrifice themselves for the good of other people, either in their families or in caring professions. It is seen as noble for a woman to put others first even to the point of erasing herself. I have too often been accused of being selfish when I put up boundaries to try to protect myself.

However, having boundaries doesn't make you a horrible person. Protecting yourself doesn't make you a horrible person. If protecting yourself means hurting someone else, that means the relationship is a toxic, zero-sum one. Usually the other person cultivated it to be that way in order to emotionally blackmail the more congressive person into doing something they would rather not do.

Psychologist Philippa Perry talks about bringing up children in a non-judgmental environment rather than in one that's all about winning and losing, which enables them better to deal with combative energy by thinking "Hmmm, interesting person" rather than fighting back.* The idea of neutralizing combative energy reminds me of the Japanese martial art aikido, which, as I understand it, involves neutralizing incoming aggressive energy (perhaps directing it into the ground) and opening up a nonviolent space for interaction instead. (I am not an expert or any kind of

---

*Robyn Wilder, "Philippa Perry: 'Listen Carefully, Parents—and Don't Despair,'" *The Guardian*, March 10, 2019.

practitioner.) The deescalation of physical violence would be a particularly valuable form of this, but deescalation of general ingressive behavior is valuable as well. This is also a way to encourage congressive behavior in others and a starting point for building new congressive structures. It does take practice, and it doesn't always work, but that can be reframed congressively as well.

Here is a small example that I have worked on for a few years. It involves someone insisting they want to discuss something with me on the phone. This might sound normal. Indeed, some popular wisdom says that it is much better or more efficient to discuss things on the phone. However, I think this assumption deserves to be challenged.

I have found that sometimes when people want to use the phone it's just because they don't want to take the time to organize their thoughts into a written form. But sometimes it's because they want to put me on the spot and pressure me into doing something I don't really want to do, like take on more work, or accept less pay, or cause some other inconvenience to myself. In such cases, this is ingressive energy that I want to neutralize. I can do it quite simply: by using email instead of a phone call. Really ingressive people can get very upset about this, and I sometimes wonder if it's because they know (consciously or unconsciously) that I have leveled the playing field where they had hoped to have an unfair ingressive advantage. I took my inspiration from a wonderfully congressive vice principal of a school who commissioned some work from me but specifically suggested discussing terms by email, observing that it can be very uncomfortable discussing them in person. It is often said that women don't like asking for pay raises, but I think that congressive people of any gender probably don't like asking for pay raises, or for anything at all. I think one way to make that act require less ingressivity from me is to do it by email instead of in person.

Every time I feel ingressive energy coming at me (which is often) I take it as an opportunity to practice this process of neutralizing it congressively. Here is a picture of how I view the options when faced with incoming ingressive energy:

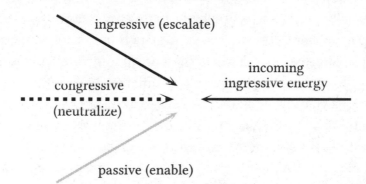

I find it good practice to try to think up three types of response, as in the preceding diagram. I also find it instructive to imagine what is the most immediate response that would pop into my head, and also what is the response that I would previously have fantasized about when trying to come up with acerbic put-downs. In the Appendix I've given some real examples of ingressive things that people have said to me, and three types of response I have thought of.

I have found that this is a subtle way to shift the balance of power in an interaction with someone behaving ingressively. As men currently hold more power in society, structurally, it is unfortunately typically men who try to wield ingressive power over me. But I have found that they respond much better to my congressive neutralizing of that power than if I try to wield ingressive power back.

It is hard to implement this practice of congressively neutralizing ingressive power, because of years spent jumping to ingressive responses, but there is no risk involved if you frame your effort as having no chance of failure: if your aim is to neutralize the ingressive energy, then it is possible you will fail, but if you reframe this congressively as aiming to *get some practice* at neutralizing ingressive energy and build up a few more case studies, then there is not really a concept of failure. No matter what the outcome of the interaction is, you can go away and think up different types of response later. Typically we all ingressively go away afterward and try to think up the best acerbic put-downs that we can, and then wish we could have thought of them at the time. But I have come to realize that I would rather think of ways to neutralize the ingressivity than come up with a scathing retort.

When we neutralize ingressive energy we achieve several things. We open up a space for more congressive interaction, but we also move away from rewarding ingressive behavior and move toward encouraging congressive behavior. And, crucially, we change the balance of power away from the status quo. As the status quo currently favors men, I consider every act that tries to shift that power to be an act of feminism, no matter how small it is and no matter how well it seems to go. Just as individuals can be contributing to the oppression of women even if they're not specifically misogynist, we can also contribute to addressing gender inequality even if we're not specifically helping women. I truly believe this is the power of congressivity.

This is a good start toward our ultimate aim, which is bigger than just learning to be more congressive ourselves and encouraging congressive behavior in others: it is to create a more congressive and thus more inclusive world. This is the subject of Chapter 7, the last chapter.

CHAPTER 7
DREAMS FOR THE FUTURE

HERE IS MY dream train design:

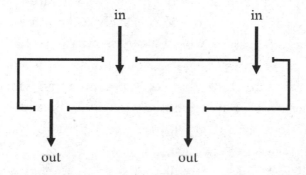

Every train platform would use both sides of the train: one side for entering and the other side for leaving. This solves the problem of people crowding around the doors and not letting people off before trying to get on. Offsetting the doors on either side solves the problem of people not moving down to the middle of the car because they're so determined to be near the door for getting off. This is a tiny microcosm of a structure that has congressivity built into it so that the advantage of ingressive behavior is neutralized. This particular example is mostly a fantasy, as it would be ludicrously expensive to rebuild all existing train stations and trains to this design, although situations with entrances and exits on opposite sides do exist on a small scale, such as on some light rail facilities at airports for transfer between terminals and on some elevators, often at underground train stations. I admit to feeling slightly gleeful when I get in an elevator where I know the exit is going to be on the other side and I watch inconsiderate ingressive people crowd around the entrance without realizing that not only are they inconveniencing everyone else but they're inconveniencing themselves as well.

I'm going to propose some dreams for the future, ways we could restructure the whole of society to be more congressive. In some cases I will build from situations that already have congressive possibilities, or perhaps from small-scale congressive structures that could become more widespread. In other cases I will dream freely to help us break free of our assumptions, even if, as with the trains, practical implementation is harder to envisage.

This is my answer to all that is written about feminism that says we need to change the entire system but doesn't suggest how we could do that. I think we can address this, equipped with our new ideas from the second part of the book. In particular, I think we can address the problems described in the first part, which had to do with still being stuck in one-dimensional thinking along lines of gender. We were stuck thinking along the lines of men and women being "different," and distracted by questions of whether those differences are innate or not. So we were stuck making a false choice between various gendered ideas of how to change the status quo.

We can now move away from all the gender-based choices. We can move away from pseudo-feminism or "leaning in," in which women are exhorted to become more like men in order to be successful. We can also move away from the opposite, in which men are asked to become more like women. We can move away from "reverse sexism," in which women are deliberately favored to make up for past oppression, and away from anti-feminism, in which women are told they simply biologically don't have the characteristics to be successful.

Instead, with the ideas of ingressive and congressive behavior we can free ourselves from that one-dimensional trap and consider how to make more congressive structures in society, which will stop us from favoring counterproductive ingressive behavior. I believe that this in turn will bring out the even greater potential of more congressive people and encourage everyone to be more congressive, which will in turn strengthen the congressive aspects of our new structures.

Many mathematical advances come from making tiny increments that build together into enormous wholes. If we try to understand or change the whole structure at once, that is indeed daunting, but I believe that if we seek to understand and change small parts and build those small parts up into large ones, we can make a difference, as I have already done in various aspects of my life at many levels, including personally, interpersonally, and professionally. Pure math is a place of dreams. It's about dreaming up new concepts and new structures. Sometimes it's about dreaming up new concepts that generate new structures, like with imaginary numbers. Sometimes it's about dreaming up new structures for old concepts, like category theory.

Unfortunately, math too often seems like a place of rules. It can seem like a place of rigidity and constraint. While that is true at some level, there is a deeper level that becomes increasingly important as you get further into research. The deeper level is about understanding the *point* of the rules and constraints. If you understand exactly what the nature of the constraints is, then you can move more freely. If you understand exactly what the rules were there for, then you can dream up productive ways to break those rules and create better worlds.

In Chapter 6 we discussed ways in which individuals could become more congressive, even within ingressive systems. But if we can collaborate to make more congressive systems, then it won't matter that some people are still very ingressive, as the system will dissipate that energy. And I think we can make a virtuous circle, as individuals influence the structures and the structures in turn influence more individuals.

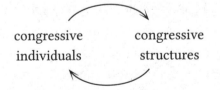

congressive individuals

congressive structures

This is a step further than just neutralizing ingressive behavior.

One abstract example of this is the "prisoner's dilemma," which I (and many others) have written about elsewhere.

It's a favorite thought experiment in philosophy and game theory showing that collaboration can produce better results than competition. In this hypothetical situation two prisoners are given the opportunity to denounce each other. If neither of them does, then they will be sentenced to, say, two years each. If they both denounce each other, they will get five years each. But if just one of them denounces the other, the denouncer will go free and the person who stays loyal will get ten years.

If the prisoners are not allowed to collaborate with each other, this is tricky. If the other person denounces you, it would be much better also to have denounced them, so that you get five years instead of ten. If the other person doesn't denounce you, then it would still be better to have denounced them, because you'll go free instead of getting two years. So if each person acts as an individual, they will denounce each other and they will each get five years; this would count as "logical" behavior. But it is ingressive logic. Thinking congressively, they could both stay loyal and end up with a better outcome: only two years each. This could work even if they're not allowed to communicate with each other to collaborate, as long as they both think congressively and trust the other to think congressively as well.

This structure still relies on trust, but I like dreaming up structures that automatically induce congressive behavior even from people who might resist it, as with my dream train design. I believe that this is a powerfully congressive way in which we can start to change the power structures of the status quo. And since those power structures currently still favor men, I think this is a form of feminism. This might sound gendered, when the whole point of this book is to argue for an ungendered approach, but I think this is an ungendered approach to feminism.

Just as there is indirect bias built into the system even when it doesn't explicitly discriminate against women, I think we can reverse that without explicitly working along gendered lines either. This really is using our new dimension to escape the traps of our old one.

I will start with areas I have the most experience with—math, in terms of teaching and research, and education more broadly. But the dreams can expand and be as big as anyone wants.

CONGRESSIVE MATH

As we've discussed, math is still a very problematic area when it comes to gender bias. We need to work to overcome explicit bias against girls and women in the form of prejudice, stereotype threat, and the lack of role models, but we also need to work to overcome indirect bias that might be happening ingressively through the way math is currently being taught at all levels. A more congressive approach directly addresses the latter issue, but I think it could address explicit prejudice as well. If we focus on the value of congressive aspects of math, we might notice that the types of people we have previously associated with mathematical brilliance (for example, men who talk themselves up a lot) are really being overvalued.

Congressive math is what I have been teaching to art students for several years. I think it's also what is introduced to children at the very beginning of school, when it's all about play and exploration with blocks and toys and other things they can touch and feel. It comes back around to being like this at a research level, but by that time I think we've already put off far too many congressive people with the phase in between.

I think math should be congressive all the way through. We really don't need to train people to be human calculators anymore, now that we have actual calculators with us more or less all the time (for example, on our phones). So math could be more congressive by being about exploration and processes. It could be more about ways of thinking than about knowledge.

I would like to see a non-cumulative curriculum so that each stage doesn't depend on the previous stage. The traditional model is more like a series of hurdles that get higher and higher and are specially designed to weed people out at each level. Not only is this ingressive, it's also counterproductive, as we are not weeding out the right people.

I would like to see category theory taught in high school, as it has great potential for showing math as a way of thinking rather than merely an exercise in getting the right answer. But congressive math is not just about the subjects we teach; it's also about the *way* we teach. It should be more about depth than speed, about invention and growth rather than finding the "math people" and separating them from the "non-math people."

Some people might worry that if we move to a congressive way of teaching math, it will be more inclusive but with the side effect of lowering standards. Of course, this is often a criticism of the idea of getting more women into math—some people believe that women really are worse at math and so to get more women into math we'd just have to lower the standards. But I think the example from Finland shows that this does not have to be the case.

Moreover, the "standards" in question are also ingressive standards. If we focus on what we can get people to achieve, usually as tested by some set of exams, then we are aiming for something dangerously limited that is unlikely to have a lasting effect on students' lives. A more congressive approach is to focus on students' experiences of math and make sure that whatever happens they maintain their interest, curiosity, and appreciation for it. If we train them to be able to spew out facts in an exam but hate math, then as soon as the exam is over they will purge all that they have learned from their brain and feel negative toward math forever, in which case I don't think we have achieved much. In fact, we've achieved something negative; we'd have done better to teach no math at all.

Whereas if we aim to nurture and strengthen interest and appreciation, then we are more likely to nurture adults who appreciate math rather than spurn it. We need more focus on this congressive aim. A gratifyingly common response from audience members when I give a public talk or interview is "I wish you'd been my math teacher at school." It is almost always women who tell me this. The people who boast to me about their mathematical knowledge are almost always men. But again, it's not all men and it's not all women. To me this shows that ungendered terminology can both help us escape those divisive sweeping statements and help with the gender imbalances that currently exist.

Moreover, I think it will be better for research. Ingressive math may raise the standards for ingressive exams, but not for research, which is much more congressive.

CONGRESSIVE RESEARCH

The world of academic research has some very ingressive aspects to it, as we have already discussed, and it is even more male-dominated than earlier levels, especially in math and science. But research is a deeply congressive activity. Using ingressive filters to select people for research is thus misguided. If we nurture congressive students, then I think we can find far more research potential in everyone. And if women have previously been disproportionately put off by the ingressive environment, this approach will address the gender imbalance. Furthermore, if it really is true that women are put off by ingressivity, then we *cannot* effectively address the gender imbalance without addressing the ingressivity. Trying to do so would amount to diversity without inclusion, miserable women (like me in my academic job), and the unfortunate possibility of those unhappy women not being able to work well in the environment and so appearing to confirm prejudiced people's view that women are not cut out for an academic career.

So I hope that the research *environment* can also become more congressive. It has been moving in that direction in a promising way, especially with the help of technology. Some very heartening congressive aspects, at least in math, are the rise of global collaboration and the free sharing of both research and exposition, as we discussed in Chapter 5. This is better than the old-style ingressive research based on secretive competition, a race to be first, and publication in expensive "prestigious" journals.

The system for publishing research is an example of a structure that can push people to become more ingressive or nurture people to be more congressive, depending on how it is set up. An example of a congressive version is the "registered reports" format for publishing research.* Unlike the ingressive system, where dramatic positive results are favored, in this congressive system peer review is conducted on the research question and the methodology *before* the data are collected. Thus it is the process that is reviewed, not the answer. The idea is that if the question is interesting and the methodology is sound, then the result of the experiment is interesting whether it turns out to be positive or negative. Psychology researcher Alexander Danvers writes that when this method is used "there are no research failures."** He argues that the (ingressive) approach of trying to get "eye-catching, wild" results does not do as much for cumulative scientific understanding, and that celebrating such high-risk work when it does go well skews scientific culture so that

---

*"Registered Reports: Peer Review Before Results Are Known to Align Scientific Values and Practices," Center for Open Science, https://cos.io/rr.

**Alexander Danvers, "Addressing the Academic Arms Race," *Psychology Today*, August 30, 2019.

scientists spend more time on questions that are of interest only when the results are positive. The ingressive approach also leads to publishing results with low statistical significance and no explanation, and to publication bias, where "failed" experiments are not reported because only "successful" ones are of interest.

Actually, my radically congressive vision for a better research world would abolish the whole idea of publishing in journals. It might seem that peer review for journal publication is democratic and thus congressive, but it's a lot more ingressive than it sounds. Peer review is too often more like gatekeeper review than like democracy, more about exclusivity than about dissemination. I would rather lessen the power of the gatekeepers. This could include sharing all research freely and having a democratic review process to evaluate it on a scale, rather than the all-or-nothing judgment of publication or rejection. We could also share grant money widely among more people rather than making everyone compete for a very small number of huge grants. We could allow prizes to be shared between more people, and reduce the hold of "superstar" culture and the adulation of supposed "geniuses"; this would allow us to congressively recognize that many people do many different types of helpful research, rather than ingressively celebrate a few and ignore everyone else.

The fixation on exclusivity is at the root of many of these problems. It is the ingressive idea that keeping people out is a sign of quality and that letting people in is a sign of weakness. It is related to the false dichotomy between research and teaching, in which people think breaking new ground is more important than making it possible for people to come with you. It is also related to a culture of experts who get their self-esteem from being able to do things that others can't do. A congressive research world would value teaching and public communication. Congressive experts get self-esteem from helping others understand what they do, and they remember to teach not just proficiency and expertise but love of the subject, love of learning, and love of curiosity.

CONGRESSIVE EDUCATION

I have already described how I have made a congressive classroom where everyone is contributing to the collective learning and ingressive behavior provides no advantage. I am lucky to teach in a very congressive institution, an art school. Perhaps that is a more obvious environment to be congressive, as art is more obviously congressive than other subjects, but I think we can take inspiration from it to see how the education system at large could become more congressive.

The ingressive system seems to be driven by the idea of assessment, so perhaps we need to start by dreaming of a congressive system to replace that. Art is obviously much less about right and wrong answers, and assessment is likely to be based on a portfolio rather than an exam. This has fewer pitfalls in art than if we tried to have portfolios for, say, mathematics or essays. It is harder to pay someone online to produce your art for you (unless it's digital art), but it's rather easy to pay someone to write your essays for you. There are even online services where you can pay math PhD students in distant countries to do your homework for you.

There are some more congressive systems of assessment by "standards" or "descriptors" rather than by grades, where instead of a single ranking there is a description of various important aspects of skill in certain areas, and the aim is to achieve some proficiency in each one. This encourages us to think about what skills education is really about and aim to teach proficiency in those instead of just aiming to achieve certain grades.

Those skills might include abstract ones like making, following, and analyzing logical arguments; expressing thoughts clearly and compellingly in different forms; organizing, assessing, and assimilating new information; making connections between situations; and providing a range of different approaches and points of view.

Then there might be more concrete life skills such as languages, writing and communicating aimed at different audiences, presenting in public, computer programming, cooking (and other DIY and craft skills), budgeting and investing, fact-checking and debunking news stories, awareness of history and other cultures.

I would add some important human skills such as kindness, generosity, self-awareness, empathy, listening and validating, not breaking people's hearts. Admittedly that list reflects my desire for more congressivity. It's possible that some aspects of my dream congressive education exist in very progressive schools or progressive education systems, possibly in Finland. I also have a list of "pet peeves" of human interaction that I sometimes wish were addressed in school: how to speak at a volume so that the people you're talking to can hear you and nobody else can (which includes learning both to project and also lower your voice), how to avoid hitting "reply all" to a group email, how to be on time for appointments and meet deadlines, how to walk through a door and make sure you don't let it close on the person behind you.

Those are, of course, big dreams that can't be changed overnight. However, one small area that is part of education but can be directly and quickly influenced is any sort of question time.

CONGRESSIVE QUESTION TIME

One specific area I have developed in a congressive way is question time after a talk. Typically questions are solicited by asking people to raise their hands, and then the questions are taken on a first-come-first-served basis. Not only is the selection process ingressive, but the whole situation of getting people to ask questions out loud in front of an audience is extremely ingressive. It does not encourage congressive questions, which would come from genuine curiosity. Instead it encourages ingressive "performance"-style questions from ingressive people who have no fear and whose main aim is often to try to demonstrate what they perceive as their superiority to the speaker and everyone else in the room. Typically the much more interesting exploratory questions come from congressive people who ask quietly afterward rather than asking in front of everyone.

In an ingressive question period, questions typically only come from white men. In case you're wondering if I'm suffering from confirmation bias: on one occasion, at the

Auckland Writers' Festival Waituhi o Tāmaki, people with questions queued up at a microphone, and after a few questions a white man shouted from the audience, "Why are all the questions from white men?"

I have tried addressing this by asking for the first question to be from a woman, as some studies have shown that this results in more women asking questions in the rest of the session. But all that happened when I tried this was that one man got so angry he took the microphone and shouted angrily at me for minutes on end. (The microphone was eventually taken away, but he continued to yell.) And asking for the first question to be from a woman is an example of "reverse sexism" in the small microcosm of a question time. (Although I stress that reverse sexism doesn't actually exist according to the contemporary definition of sexism as only being about the group in power being prejudiced against the group not holding power.)

It is less divisive to degender our approach and think of a congressive version of question time instead, using this new dimension. Some proposals include having people write questions on notecards or on an online forum, but in my experience this is like antibiotics that kill off the standard bug and leave the most resistant strains of superbugs behind: the mildly ingressive people are neutralized, but the really ingressive people are actually emboldened by the anonymity of the situation. Furthermore, the congressive aspect of human interaction is removed, as I can't look the asker in the eye or understand anything instinctively about why they are asking such a question. It leaves the ingressive energy bouncing around the room.

Eventually I devised a way to run a congressive question time that neutralizes ingressive behavior without falling into the one-dimensional traps. Now instead of asking for raised hands I invite everyone to have discussions with those around them. I walk around the room hearing what people would like to explore, then return to the front and share those thoughts with everyone along with my reflections. Not only does this open possibilities for congressive people to participate without pressure or fear, but when the potential for performance and a public assertion of superiority is transformed into a human interaction, then somehow the ingressive behavior dissipates. As it happens, it also means I interact with far more women than in a normal ingressive question time, in which I hardly ever see any women ask questions. Sometimes the non-men in the audience just want to tell me about a part of the talk that they really appreciated. This gets filtered out in normal question times, especially when the moderator sternly reminds the audience that "questions end with a question mark." This well-meaning attempt to prevent ingressive ranting typically just means that the ingressive ranters rant away and then end by saying ". . . and my question is: Have you thought about that?" and the congressive expressions of appreciation are lost.

CONGRESSIVE WORKPLACE

I realize that in the wider world of work some ingressive structures might be more necessary than in academia, but we can still dream. Promotion might be necessary because people gradually take on more and more responsibility. One way to make it more congressive might be to make it more about sponsorship and mentorship than about putting yourself forward. My career was substantially and crucially facilitated by various people supporting me along the way, rather than by me putting myself forward—at least, that's how it felt to me. (Incidentally, all the people who put me forward in those career progressions were men.) We do have to make sure this doesn't just sound like old-boys networks. In *Inclusive Leadership*, authors Charlotte Sweeney and Fleur Bothwick warn of the perils of creating support networks that too quickly become exclusive in turn, as can happen—alas—whenever excluded people get together and emulate their excluders by excluding others in turn.

It is also important that this should not be about some sort of arbitrary "sisterhood." Women's groups can be off-putting to some women and also seem exclusionary of men. They also too often neglect intersectionality and just end up creating a new hierarchy of power inside them, with straight middle-class able-bodied cisgender white women at the top, imposing their power on all the other women. Instead we could make congressive groups, for congressive people and people interested in congressivity.

Another aspect of a congressive workplace would be flexibility, so that everyone can play to their own strengths rather than conform to an imposed structure. This is a way to get much more out of everyone, as we've seen in the case of Dame Stephanie Shirley, and yet in my institutional academic job allowing people to play to their strengths was seen as unfair.

In *Becoming*, Michelle Obama writes of working to remedy the imbalance in her law firm, which predominantly hired people who were male and white. She had to persuade the recruiting team to look beyond the usual metrics of prestigious universities and exam results and consider the applicants' backgrounds to understand whether they'd coasted on privilege or raised themselves up from difficult beginnings. I would now call that thinking congressively rather than ingressively about hiring.

Different industries have very different characters as driven by their particular fields, but another one that I know quite well, other than math and education, is music.

MUSIC

Music and creative arts are fundamentally congressive, as they are about creating things rather than about right-or-wrong situations. When musicians get together to make music in a group they are not trying to beat each other. The goal is to make wonderful music together, and that is achieved by everyone contributing and supporting each other.

Unfortunately, music is another part of the world that has become rather an ingressive industry, with competitions, stressful auditions, the pressure of performing onstage, critics making ingressive judgments, prizes, and superstar idolization. In a music competition you contrive scarcity by declaring that only one person can win. Whenever there is a move away from this in which everyone gets a prize, there is an ingressive backlash from people who think that the "everyone gets a trophy" culture is ridiculous and doesn't breed excellence. The idea that competition produces better results is ingressive, and people who do actually achieve more in competitive situations are more ingressive. The stress of a competition means that you have to have a certain kind of emotional disposition to withstand it, and, as Alfie Kohn points out in *No Contest*, that might be the exact opposite of the emotional disposition needed to play music sensitively.*

By contrast, the idea that a supportive and collaborative atmosphere produces better results is congressive, and congressive people may well achieve more in that type of atmosphere. I have founded a consciously congressive music organization in Chicago called the Liederstube, focused on classical song. There is no barrier to entry—anyone can come and sing, and anyone can come and listen. There is also no barrier between the performers and the audience—everyone sits together in a cozy room with an atmosphere that is more like a party than a concert. It is less like performing (which is ingressive) and more like sharing music we love (which is congressive). There is no emphasis on perfection or correctness (which is ingressive). The pressure of performance on a stage is completely removed, and this can result in more emotionally touching music. But it's also more efficient, because when you're preparing for a performance a huge proportion of the work goes into preparing to withstand stressful conditions.

So for a tiny fraction of the effort of preparing a formal concert performance we can make music that is almost as good except that maybe we play more wrong notes. It's

---

* *No Contest* quotes critic Will Crutchfield on this subject.

rather ingressive to care about wrong notes anyway; it is more congressive to care about the feelings we communicate and share with everyone in the room. I like the congressive version and think it draws some people into the classical music world who are put off by the typical ingressivity (and the ticket prices).

I would like to run "auditions" like this, where it's less like a competition and more like getting to know each other in a genuinely congressive music-making environment. We still find out what we need to know about each other as musicians. It takes a bit longer, but it filters congressively for something that should be congressive.

I invite and urge you to think about any other aspects of life in this way—how ingressive they are, how ingressive they really need to be, and in what ways they could actually benefit from being more congressive. And then I urge you to dream up ways to nudge them in the direction of more congressive behavior. This doesn't have to be in daunting large structures such as entire industries—it can start with the structures of our personal interactions, such as in any discussions we have with other people.

CONGRESSIVE DISCUSSIONS

Discussions are a small form of social structure, the smallest being a discussion between just two people, or even a discussion with yourself. Too many discussions are really arguments aiming to establish superiority or dominance. Congressive discussions are instead a way to establish a connection between people as well as greater understanding. Perhaps women tend to do more of the connecting and understanding and so have more meaningful friendships; of course, referring to gender in that way is likely to be divisive, so it's better to say that congressive behavior fosters more meaningful friendships and everyone can learn it. (I've given some specific tips on this in the Appendix.)

Unfortunately, along with our general obsession with competitions and "winning," we are very attached to the idea of winning arguments. In fact, when I do interviews about my previous book, *The Art of Logic*, the headline for the resulting piece often looks something like "Abstract Mathematics Helps Us Win Arguments" despite the fact that one of my major points in the book is that I think it is more productive to seek to understand each other than to "win."

We see copious examples of ingressive arguments, including formal debates in debate clubs, during elections, and in political slots on news programs. When politicians are interviewed on air the situation is usually one of mutual attack, but I often find that in that process I don't discover anything more about what the politician thinks or why they think it. Moreover, the politician is rarely genuinely challenged about their position, as I think they would be if they were asked to expand on it and talk about it in depth rather than rapidly parry sharp blows from the interviewer.

In non-political interviews there is often a more congressive approach, not based fundamentally in attack. If a scientist has made a breakthrough worthy of news, then the interviewer's aim is more obviously to get the scientist to explain it to everyone (although journalists who don't specialize in science do sometimes have a skeptical "What on earth is the point of this?" spin on more esoteric aspects of science).

What if we had congressive political interviews as well? Perhaps we'd learn more and also have more real challenges to our politicians instead of the usual situation where they can just spew the same soundbites over and over again or deflect questions by answering something else entirely.

We could learn more from congressive discussions in our daily lives too, instead of having divisive, ingressive arguments that don't help us make progress. For example, the so-called oppression olympics take place anytime people in a discussion are competing with each other to show that they are the "most oppressed." Too often in the environment of identity politics it seems that only the most oppressed person in the room is allowed to have experiences that are considered valid, because everyone else counts as "privileged." This is sometimes taken as an argument showing that identity politics is unhelpful, but I think it only shows that *ingressive* identity politics is unhelpful. If we acknowledge everyone's privilege and everyone's misfortune and everyone's relationship to everyone else, then we can work congressively to reduce *all* oppression instead of being distracted by the oppression olympics.

During ingressive arguments I have rarely seen anyone persuaded of anything. We can retrain ourselves to make arguments more congressive by seeking to understand things rather than defeat other people. In those situations you might suddenly find that you don't mind discovering that you're wrong after all, and that you learn a lot more about something. I have learned about congressive arguments from the most congressive person I know, my friend Gregory, with whom discussions are always about discovery and not about right and wrong. I always learn from those discussions and never feel bad; rather, I feel that progress has been made. It might be particularly hard between a woman and a more ingressive man because of the gendered assumptions that still persist in society about who is supposed to play the role of an "expert." It still frequently occurs that my very existence as a woman mathematician makes a man feel so insecure that he immediately starts trying to belittle me. Sometimes for my psychological well-being I don't bother engaging, but sometimes I can concentrate on taking a congressive approach and then have a more productive discussion in which I seek to understand where his insecurity is coming from and in the process he understands more as well, once the ingressive distraction is removed.

And remember that ingressive distraction just benefits the people in power, currently dominated by men, and contributes to the oppression of women and others. Competition obstructs collaboration, and ingressive energy obstructs us from working congressively to change the power structures. It might seem that there's little that an individual can do under those circumstances, but here is a very inspiring story of one woman who effected powerful congressive change, starting small.

RESTORATIVE VERSUS PUNITIVE JUSTICE

Becoming Ms. Burton is an extraordinary and important memoir of an extraordinary woman, Susan Burton, and her journey to becoming a social activist rescuing women from a systemically racist criminal justice system in which she was herself trapped for years.

The first part of her story is tragedy upon tragedy. She was abused and raped as a child and teenager and spiraled into worse and worse circumstances, like so many black people in the United States who are neglected and exploited on all sides. Her five-year-old son was killed in a hit-and-run by a police officer, and in her grief she fell prey to the crack cocaine wars that were just beginning to rage in LA. There began fourteen years of her bouncing in and out of prison for drug offenses, victim of a justice system that sought only to punish women like her, not to help them.

Eventually she managed to free herself with the help of a residential rehab program, and she turned her attention to helping others, first on an individual scale as a caregiver and then by setting up A New Way of Life, a nonprofit providing residences for women coming out of prison to help them help themselves. She began small, with just one house for ten women in 1998, and gradually expanded; to date A New Way of Life has helped more than a thousand women, including reuniting more than three hundred of them with their children, according to the organization's website.*

She writes compellingly of the injustice of a system that simply incarcerates people who need help. It is even senseless from a financial point of view—she writes that taxpayers

---

*A New Way of Life, "What We Do," anewwayoflife.org/what -we-do.

pay up to $60,000 per year to incarcerate a person, a sum comparable to tuition at an elite university. This sparked the slogan "Yale not jail!" A New Way of Life apparently provides its services at a third of the cost of incarceration.

Incarceration typically leads to more incarceration, as people get stuck in a loop that becomes increasingly hard to escape. She writes, "Jail had done nothing to stop my addiction. Education, hard work, dedication, a support system, and knowing there were opportunities for me and that my life had value: these were what had made all the difference." What worked was a congressive restorative approach rather than an ingressive punitive one.

An unnuanced point of view is that if you don't commit crimes, then you don't get stuck in this loop in the first place. However, Ms. Burton's book sets out quite starkly the two-tier system in which white people and black people are treated differently for the same minor offenses. Moreover, 85 percent of women in prison are victims of physical or sexual abuse, and these women are "disproportionately black and poor." Many became drug addicts in desperate circumstances, manipulated and exploited by the dealers, but they are treated only as perpetrators, not victims.

A more congressive point of view says that if we helped people more in the first place, it would be a much more effective way to reduce crime, and that punishment doesn't reduce these issues as much as help and rehabilitation do. This might seem like a naively utopian dream for the future, but congressive social structures have existed for a long time, just not in white capitalist societies.

NATIVE AMERICAN CULTURES

In *The Sacred Hoop*, Native American writer Paula Gunn Allen writes about tribal social structures that were deliberately undermined by white colonial invaders. She describes tribal societies as egalitarian, organizing events and phenomena not by hierarchies, as in white (Christian) societies, but with everything related to everything else in a harmonious or unified way. She even writes of traditional war that it was "not practiced as a matter of conquest or opposition to enemies" but instead was a ritual to gain the attention of supernatural powers.

She describes Paul Le Jeune leading the Jesuits to try to convert the Montagnais to Christianity, writing, "How could they understand tyranny and respect it unless they wielded it upon each other and experienced it at each other's hands?" Le Jeune was determined to change the minds of these people who "did not punish children, encouraged women in independence and decision making, and had a horror of authority imposed from without," and who did not believe in people assuming superiority over other people: congressive values. She describes the development of tribes over five hundred years of colonization as they progressively moved from gynocentric egalitarian social systems toward the European patriarchal, hierarchical style. I would say: from congressive to ingressive.

Allen writes, "Under patriarchy men are given power only if they use it in ways that are congruent with the authoritarian, punitive model." I would extend this to say that in an ingressive society people are given power only if they use it

in ingressive ways. As we have discussed, this has encouraged everyone of all genders to be as ingressive as possible
in our ingressive society; alternatively, it has caused those
who don't want to be ingressive to give up and resign themselves to being powerless, or to latch on to some borrowed
ingressive power from someone else (for example, by mar-
rying a powerful man). But Allen shows that it was not like
that before the arrival of the white colonizers. It appears that
congressive society isn't something new we need to build but
something old we need to restore.

CONGRESSIVE DEMOCRACY

Democracy is a fundamental way in which we interact with
the ingressive power structures of society that the European
colonizers spread around the world. Any form of democracy
is of course more congressive than a dictatorship, but different nuances of ingressivity and congressivity occur within
the different types of democracy that exist around the world,
both in how the democratic systems operate and in the election processes used to select politicians. I think our systems
of power are still predominantly ingressive. Of course, politics around the world is still largely dominated by men, and
so this is another area in which a more congressive approach
could change that status quo. What might that mean?

Considering systems of democracy, direct election of a
president (as in the United States) is much more ingressive,
whereas in other systems such as the United Kingdom's the
head of the government is just the leader of whichever party
has the most seats. It is possible that this is why it is taking

much longer to get a woman president of the United States than a woman prime minister of the United Kingdom or a woman chancellor of Germany.* But I would prefer to say that really it makes it harder to get a congressive leader.

Whatever the system, election campaigns are increasingly ingressive. It's true that there is genuine competition involved, as there is a limit to the number of people we can elect, thus the scarcity is real. However, there are more and less congressive ways of going about this selection process, as can be seen by looking at different systems across the world and across different times in history.

Such systems are perhaps more congressive in countries other than the United States—for example, in the United Kingdom, where spending is limited by law, and where negative campaigning is less prevalent (although that seems to be changing for the worse everywhere). Election debates have become extremely ingressive. There is a video still available of a US primary debate in 1980 between Republican candidates George H. W. Bush and Ronald Reagan in which they essentially agree about wanting a compassionate approach to immigration from Mexico. This sort of polite, compassionate consensus seems unthinkable now, especially on that topic.**

---

*See, for example, Zack Beauchamp, "The US Has a Female Presidential Nominee for the Frst Time. Here's Why It Took So Long," *Vox*, July 26, 2016.

**"George H. W. Bush and Ronald Reagan Debate on Immigration in 1980," YouTube, posted by *Time*, February 3, 2017, www.youtube.com/watch?v=YsmgPp_nlok.

A two-party system is another ingressive aspect, as is the first-past-the-post (FPTP) voting system (also called "winner-take-all," which makes it sound even more ingressive). In fact, FPTP voting reduces the impact of small parties, so it tends to cause a de facto two-party system. It's ingressive because it's basically like a zero-sum game—to win more votes you have to make someone else lose them (unless you can rely on motivating substantial numbers of previously apathetic people to vote).

Voting systems are a tested example of how a more congressive structure can affect people's behavior and influence them in a different direction, which is just the sort of thing I believe we should dream up and implement more widely.

Voting systems that are more congressive than winner-take-all exist, such as proportional representation, systems with multiple rounds, or ranked choice. In these more congressive systems votes for smaller parties aren't so obviously wasted, as they can still count toward a proportion of seats, or in a system with multiple rounds candidates who are knocked out will have their votes redistributed according to the rankings selected by voters. One such system that is already in use in some parts of the world is alternative vote, also known as single transferrable vote, ranked-choice voting, or instant runoff. The idea is that if there are many candidates, then the one with the fewest first-choice votes is eliminated, as if in a multi-round knockout challenge, and then everyone votes again. To avoid the logistical complication of everyone *actually* voting again, voters are allowed to rank all the candidates on their ballot in their order of preference, so that if their top choice is knocked out, their vote can then be redistributed to their next choice. This process of knocking people out and redistributing votes continues until someone has a majority.

Some people object to ranked voting systems on ingressive grounds, claiming that it means "the winner doesn't necessarily win." However, this is only true if you define the winner to be the person with the most first-choice votes. So all that the ingressive person is objecting to when they say "the winner doesn't necessarily win" is that winner-takes-all and ranked voting might produce different results—which is the whole point.

It is hard to see how to move to any such congressive systems when the current system gives disproportionate power to certain parties. Those parties hold power and benefit from the current system, so it is particularly hard to know how to change that except by slowly influencing more people to be congressive and to believe in congressive structures. Some places in the United States have managed to adopt more congressive systems by voting them in, including the state of Maine and several cities such as San Francisco, Minneapolis, and, most recently at time of writing, New York City. The case of Maine is quite interesting: ranked-choice voting seems to have been successfully voted in as part of a reaction to the Republican candidate Paul LePage being elected governor with much less than 50 percent of the vote twice in a row, with the more moderate vote split between a Democrat and an independent. It seems LePage has called himself "Donald Trump before Donald Trump became popular."*

---

*Katharine Q. Seelye, "Paul LePage Was Saying Whatever He Wanted Before That Was a Thing," *New York Times*, August 13, 2018.

Indeed, apparently the more a Maine voter supports Trump the less they support ranked-choice voting.*

But supporting a congressive system shouldn't just be about wanting a particular person to win. It's not about manipulating the outcome to be the one you want, but about believing that a more congressive process is more representative. That said, if ingressive people will only support the system for self-serving reasons, they might be convinced to support it once they see that their own vote could be split, not just their opponents' vote. The United Kingdom voted against such electoral reform (by a substantial margin) in a 2011 referendum, but the political climate of the 2019 election meant that the votes on both left and right seemed in danger of being split—on the right between the Conservatives and the Brexit Party, and on the left between Labour and the Liberal Democrats. However, as it turned out, the Conservatives benefited from vote-splitting without suffering much from it, so the chance for all sides to support a more congressive system for self-serving reasons once again faded.

Aside from voting systems and party systems, the whole adversarial style of modern politics might be regarded as an impediment to sorting anything out. Persuading people to be less adversarial in a system that rewards it is likely to be futile, and so the key would be to change the system into one that doesn't reward such behavior. Again, it's a way we could create a congressive structure to nurture rather than force congressive behavior.

---

*Joseph Anthony, "Ranked-Choice Voting Is a Partisan Affair in Maine," *Bangor Daily News*, September 24, 2019.

This is another potential benefit of ranked voting systems: they encourage a less combative and partisan campaign because candidates need to appeal to their opponents' supporters in order to pick up second-choice votes. Such systems also increase the chance of having several smaller parties who need to cooperate with each other in order to get anything done. Another alternative system involves "constitutional conventions, cross-party forums and citizens' assemblies" to supplement parliament.* I would call this a more congressive approach to running a country. Of course, the question remains of how we change the status quo to achieve that.

CHANGING THE STATUS QUO

People naturally resist changing the status quo if the old system has worked particularly well for them, as with political parties who benefit from the winner-take-all voting system. When people have enjoyed disproportionate power, they will resist having it taken away.

Curiously, though, I have not yet met much outright objection to the notion of ingressive and congressive behavior from men; rather, men have shown a lot of enthusiastic support and sometimes relief in response to these ideas, which to me is testimony to one of the benefits of ungendering our argument. Of course, it's possible that I just don't meet many of the men who are really holding disproportionate power. However, I have encountered objections

---

*John Coakley, "Brexit Has Nearly Broken British Politics. Here's How to Fix It," *The Guardian*, March 30, 2019.

from women who have been very successful by learning to emulate ingressive behavior. Perhaps they haven't imagined another possibility or are afraid they will be worse off in the new structures.

The status quo is evidently supported by those who win in the current structures (of whom there are not that many), but unfortunately also by those who are convinced that they could win—indeed, that they might soon win. Perhaps this is an essential aspect of the "American dream," the idea (or rather, the dream) that in this structure anyone can rise up, even all the way to the top. Convincing people that they are about to rise up while simultaneously keeping them down is a particularly pernicious way to maintain power.

Unfortunately, most people who think they're about to win are deluded, like addicted gamblers, because most people do not win in an ingressive situation. But the winners benefit from those who continue to delude themselves that they can win. In the case of gambling in casinos, the winners are the casinos themselves. The casinos depend on people being convinced that they can win, but the only way to be sure of winning, unless perhaps you're a world-class poker player, is to get out of the casino and do something else.

I think a new approach based on congressive values will be better for the whole of society, including men, women, and oppressed minorities. It just won't feel better for the people who currently hold undue amounts of power and who want to cling to it. It's also important to remember that there are problems separate from this that we must still address, including problems of prejudice based explicitly on gender, race, and other identities. But the idea is to separate those out from the questions of character that we have been discussing throughout.

Aside from whether it changes gender imbalances, I believe that a congressive system is the right approach for the good of society, inclusivity, and equity. But I do think that making more congressive systems and structures could lead to better outcomes about gender imbalances, with a less divisive dialogue. It's my congressive feminist dream.

"Smash the patriarchy!" is the somewhat ingressive battle cry of some branches of feminism. I prefer to say something more congressive and less divisive: Let's transform our world into a more congressively led future. Let's reward, encourage, and nurture congressive behavior and build congressive structures. At the same time, let us stop fabricating ingressive structures like unnecessary competitions and contrived zero-sum games, and stop using ingressive filters for congressive roles.

I think this is an implicitly feminist aim. The ingressive structures and our ingressive attitudes are perpetuating power structures that favor men, so anything we do to dismantle those structures is, I believe, an act of feminism. And just as bias against women can be either explicit (specifically because of gender) or implicit (through unconscious associations or structural issues), I think dismantling it can also be done explicitly or implicitly.

Explicitly dismantling gender bias means changing those structures along one-dimensional gendered lines. This is still an important part of dismantling explicit sexism, but it will not address indirect and structural bias in the system.

I believe that the new dimension of ingressive and congressive traits can help us overcome bias that can't be addressed on the dimension of gender alone. I think this is how
we can deal with implicit bias in the system that comes from
our association of character with gender, and how we can
deal with indirect bias that comes from favoring ingressive
behavior. But I think it will also have an effect on explicit
bias, as it is a way for everyone to escape one-dimensional
gendered thinking in their heads and think more clearly
about what contributions to society we want to see. With
every individual who escapes that thinking, the hold of both
implicit and explicit gender bias will be lessened.

It is hard for congressive people to change the balance
of power while everything favors ingressive people, but we
can do it congressively by collaborating and supporting each
other. This is different from the arbitrary "sisterhood" that
some feminists think all women should be in, which is in itself a divisive idea. Many women rankled at the idea that they
"should" vote for Hillary Clinton "because she is a woman."
The divisive idea of doing something for someone "because
she is a woman" can now become the less divisive idea of
doing something for someone because they are congressive.
And I think that helping congressive people in certain ways
is not just fair but also brings the possibility of huge benefits
for everyone.

We may be losing all sorts of congressive people from further education, graduate study, research, high-level jobs, and
promotions by not giving them the external validation they
need *and deserve*. Thus we leave all of those opportunities to

the ingressive people who base their self-esteem on nothing in particular and so might well be less skilled at the task in question, less willing to improve, and less able to grow.

I hope that we will all get together and think of more and more congressive systems for the good of society. We particularly need action from those in any position of power and influence, including teachers—and especially teachers of subjects that are traditionally rather ingressive, such as science and mathematics. And we need more appreciation of traditionally congressive subjects such as the creative arts.

We need action from people running companies and hiring, to find ways to make their companies more congressive if they believe in inclusivity and getting the best out of everyone. We crucially need action from parents raising children so that the next generation need not be quite so ingressive right from the start. We could also do with some academics to conduct formal research on this from the point of view of psychology, sociology, and more.

You might think that my congressive utopia is a lovely dream but can never become reality. Well, I always say that if we declare something to be impossible, then it will become impossible, in a self-fulfilling prophecy. We don't have much to lose by trying, and who benefits if we don't? The people who benefit the most are the ingressive people who are currently in power. As Alfie Kohn says, "I would prefer to see skepticism directed at the status quo rather than employed in its service." Jessa Crispin reminds us that "breaking away from the value system and goals of the dominant culture is always going to be a dramatic, and inconvenient, act."

We don't have to change the entire world at once, and we don't have to work on all scales at once. We can start small, by making personal changes in our daily lives and interactions with individuals. We can change the way we discuss things, and we can even just stop using implicitly ingressive language; for example, we can stop using the word "competitive" just to mean "good," and we can avoid using slang phrases like "for the win" as a compliment. We can encourage congressive behavior in the people around us and practice neutralizing ingressive energy wherever we encounter it.

We can then move on to small microcosm utopias, like my congressive classroom and my music salon. This could include family units (as Philippa Perry suggests), groups of friends, clubs, small organizations, and small companies. Next we could move on to bigger institutions: bigger companies, whole schools and universities. After that we could gradually move on to social structures such as entire industries, election processes, democracies.

We started with a small step, a small shift in perspective away from the dimension of gender. We started with a few case studies, a few role models. I believe that we can gradually roll these ideas out to the rest of the world.

Thinking along the dimension of ingressive and congressive traits enables us to think more clearly and solve particular problems, as a good theory should. In fact, I feel that thinking this way has given me radical clarity about how to live my life, how to contribute to the world, and how to be a good feminist, working to change power structures while still recognizing the experiences of individuals and the range of other dimensions on which they experience advantage and disadvantage.

I think that what is at stake here is nothing less than our humanity itself. One of the major features we humans have is our congressive ability to communicate and collaborate. This is how we have formed communities at the scales we have, and made progress in ways that no other animal has. But there is also a major ingressive aspect of humans that sets us apart: that we enslave people (and other animals), exploit them, and exert power over them.

Things don't always have to be a competition, but perhaps if there is only one competition, it is between the ingressive and congressive forces in humanity. I hope that we will choose congressivity and work together, congressively, to build a better future for everyone.

POSTSCRIPT

A **WEEK INTO** our coronavirus lockdown in March 2020, I heard a compelling story from my dear friend Amaia. A poet, optimist, and believer in nature, she was looking for hope in the midst of this global crisis and found something quite unexpected: a fascinating example of social change following crisis, involving a troop of savanna baboons in Kenya, studied by Robert Sapolsky and Lisa Share.[*]

Natalie Angier writes about this research, introducing the idea of positive change coming out of a crisis: "Freak cyclones helped destroy Kublai Khan's brutal Mongolian Empire, for example, while the Black Death of the 14th century capsized the medieval theocracy and gave the Renaissance a chance to shine."[**] As a more recent case, the Second World War was an atrocity that must never be repeated, but during it women showed their worth as fully participating workers in society,

---

[*] R. M. Sapolsky and L. J. Share, "A Pacific Culture Among Wild Baboons: Its Emergence and Transmission," *PLoS Biology* 2, no. 4 (2004): e106.

[**] Natalie Angier, "No Time for Bullies: Baboons Retool Their Culture," *New York Times*, April 13, 2004.

and society was forced to acknowledge it and change at least some of its ways.

The baboon story, in short, is that an aggressive and hierarchical baboon society transformed into a nurturing and peaceful one in the space of one generation. The catalyst for the change was a crisis in which half of the males died of tuberculosis. Crucially, it was the most aggressive alpha males who died. There had been a fight with a neighboring troop over some food at a garbage dump. Only the most dominant males had been able to hold their own in the fight and get to the food, but the food turned out to be contaminated with tuberculosis bacteria.

The surviving society consisted of previously subordinate males, the females, and the young. Sapolsky and Share describe the transformed society, and I would call it congressive. It remained congressive even when the males who had survived the epidemic had died and new males from outside the troop had joined; the new males learned the congressive behavior from the existing members. The researchers found that although the society was still hierarchical, it was less aggressively so, with the dominant males less prone to taking out their aggression on the subordinate males and the females, and with more "affiliative" behavior such as mutual grooming to build community. And this new congressive community was better for everyone in terms of their levels of stress, without compromising their ability to survive as a troop.

I am absolutely not suggesting that we hope to wipe out all the ingressive people in the world with an epidemic. But perhaps, like the baboons, we can respond to a crisis with a

cultural shift away from our hierarchical, gendered, ingressive culture toward an inclusive, congressive future.

This global pandemic of 2020 might show us some ways
in which traditionally celebrated ingressive behavior really
is ineffective at best, and deadly at worst. Risk-taking and
disobeying social distancing instructions increase the spread
of infection. Basketball player Rudy Gobert publicly scoffed
at the risks, touched all the microphones at a press conference, and then became the first NBA player to test positive
for the coronavirus, resulting in the abrupt shutdown of the
season. Politicians accustomed to bold posturing have found
that you might be able to outposture a human enemy but you
can't outposture a virus. As industries have shut down one
by one and a large proportion of people's income streams has
been cut off, even right-wing conservatives have acknowledged that people need help from the government, possibly
even a universal basic income—albeit temporarily—in order
to survive. Suddenly congressive policy that was dismissed as
utopian even on the left two weeks ago is being proposed by
the right as a matter of urgency.

At more individual levels, those of us who are confined
to our homes are finding more ways to share things: ideas,
skills, expertise, our art. Arts institutions have been opening up their digital archives, streaming free concerts, making
virtual tours. Some people have been hoarding toilet paper,
but others have been hand-sewing face masks for hospitals
desperately short of supplies and overwhelmed by the influx
of patients.

Perhaps by the time you are reading this it will all be over;
I certainly hope so. I hope that by the time you are reading

this we are no longer confined to our individual homes, that we have gone back to eating and drinking together, going to concerts, visiting galleries, browsing in real physical bookstores. But I also hope that we have not forgotten this slight tilt toward congressivity that I am starting to see. Rather, I hope that we have built on it, and realized that these ideas generalize beyond a crisis, and that even when times are not so universally dire, we can think and act congressively to make life better.

Appendix

HOW TO BE MORE CONGRESSIVE

Here are some practical suggestions for how to become more congressive in your daily life. We could think of it as congressive personal training for anyone who wants to unlearn ingressive behavior and learn to be more congressive. Most of this is an exercise in gradually shifting emphasis rather than (ingressively) suddenly flipping your entire behavior. Remember that the congressive approach is to aim to keep practicing rather than aim for a particular outcome, and to acknowledge that it's an ongoing process rather than something that can be suddenly achieved. If it doesn't quite work on one occasion, that doesn't mean you failed; the only way to fail is not to try. If you at least try, then whatever happens you will have more case studies to think about for the future.

LOOK FOR SIMILARITIES BETWEEN PEOPLE AND SITUATIONS RATHER THAN DIFFERENCES

This is like the idea of seeking to empathize with other people rather than show how different you are. This might be in response to someone talking about their experiences, their tastes, their dreams, or things that are bothering them. Remember that it's not a competition. If someone talks about something bad that happened to them, it's not very congressive to "sympathize" by saying that something similar happened to you, but worse. Showing sympathy when something very bad has happened is tricky because if you say you understand, it might be invalidating to someone who feels that what they are experiencing is uniquely terrible. So seeking to validate someone's experiences is a very important aspect of congressive behavior, and this requires empathy in order to tell whether they want to know that their experience is unique or want to feel that their experience is universal. For example, with logistical frustrations it can be very calming to know that we are all sharing these frustrations as a community, and it can be very isolating to feel that nobody else is bothered by these things. Not to mention that if everyone is bothered by the same things, we're more likely to have a chance of changing them.

SEEK TO SUPPORT OTHERS RATHER THAN ADVISE THEM, UNLESS THEY SPECIFICALLY ASK FOR ADVICE

Unsolicited advice is ingressive, as it assumes a position of superiority—you are taking it upon yourself to assert that you know better than the other person does about what is going on, and you are assuming that there is something you can offer that they won't already have thought of. It also assumes that they are looking for a solution, and furthermore that a solution exists, which has the danger of invalidating the realness of their problem. Perhaps what they're really looking for is sympathy and acknowledgment that the problem really is annoying, frustrating, or upsetting. If you give them advice under those circumstances, they are quite likely to resist it and tell you why they think it won't work, or say they've tried it and escalate their expression of how bad they feel. It's like in Aesop's fable of the sun and the wind, when the wind blowing harder just makes the person pull their cloak tighter around them, whereas the sun shining induces them to take the cloak off.

You might feel that you are a problem-solver and you really want to solve that person's problem for them. One way to try to convince yourself (or an ingressive friend) to support rather than advise is to observe that the described problem isn't the actual problem that needs fixing—the "problem" is the need for validation. So if you really want to help fix the problem, providing sympathy and validation is in that sense solving the problem.

If you really think you might have something helpful to offer, a congressive way to offer it would be to say something like "Would you like to hear something that helped me?" and then be ready to stop if they say no. In the end a congressive approach is to help someone solve their own problem instead of swooping in and thinking you can fix it for them.

BUILD MUTUALLY BENEFICIAL SITUATIONS

There are too many zero-sum situations in the world, often unnecessarily, as with the contrived scarcity we discussed that causes unnecessary competitions. Even if you can't control large structural situations, we can all make changes in interpersonal situations.

Alfie Kohn discusses more abstract forms of scarcity as well, including interpersonal situations of "mutually exclusive goal attainment," in which each party involved can only achieve their goal by preventing the others from achieving theirs. That is an ingressive situation. This often happens in manipulative or abusive relationships, where the abuser deliberately creates a zero-sum game, convincing the other person that the only way to make their partner happy is to be hurt themselves.

To be more congressive we can build situations for mutually beneficial goal attainment instead. Even ordinary conversations have potential for this, as they often turn into zero-sum games unnecessarily, with a limited resource of "being right." If a conversation is instead about gaining understanding, then, as with education, the resource is not limited. In fact, more understanding can be gained if everyone collaborates and shares the effort rather than obstructing everyone else's efforts. Instead of a zero-sum game we have one where the outcome is "more than the sum of its parts." It is no longer an ingressive win/lose situation at all but a congressive mutually beneficial situation. Other mutually beneficial situations include playing to one's strengths in relationships (romantic, professional, or just between friends) so that everyone contributes what they're good at.

Better still, we can resist and actively dismantle zero-sum situations that other people create. This might seem ingressive, but there are congressive ways to do it. Sometimes I manage it by convincing myself that every time I defuse an ingressive situation I'm helping the world. I also remind myself that the zero-sum situation was not my fault in the first place, so resisting it need not be considered my fault either.

ADOPT THE PRINCIPLE OF CHARITY

This means seeking the most generous interpretation of someone's point rather than an antagonistic simplification. The latter is usually a precursor to a straw-person argument. This is part of a broader point of trying to understand what someone really means rather than jumping down their throat and immediately trying to demonstrate that they're wrong. We can always look for the sense in which people are right rather than pointing out ways in which they're wrong.

For example, if your partner says, "You never do the dishes!" you might immediately point out that one time three months ago when they were sick in bed and you did the dishes. That would be an ingressive response. A congressive response would be to observe that they probably don't mean that you literally never do the dishes but rather that they're frustrated by how little you do the dishes. You might then think about whether it's worth doing more dishes to help them feel better.

This is a good congressive approach whenever somebody makes a point that riles you up. This can happen on both sides of a disagreement about bias and oppression. If someone expresses frustration about "men" or "white people," try to understand that they don't necessarily mean all men or all white people, but think about the structures built into society and consider whether you can see any truth in the fact that men and white people hold a lot of power in society. This is relevant in any discussion with someone more disadvantaged than you in some way. On the other hand, if you are the more disadvantaged person expressing frustration and the man or white person (or white man) gets cross with you for being "prejudiced right back," you could congressively try to think about why they are feeling that way. It might just be coming from a selfish fear that if everyone were treated fairly, then they themselves would not have so much power. But if you have the energy, perhaps try to see that their frustration is coming from their own feeling of exclusion from power in society on the grounds of social status, wealth, sexual orientation, or something else.

CONGRESSIVE ROLE PLAY

Here are some real examples of ingressive things that people have said to me, and three types of response I have thought of, as I described with a diagram at the end of Chapter 6. This is something you can always try doing after an interaction has occurred if you find yourself regretting how it went. The more you practice thinking up congressive responses slowly by yourself, the easier it will become to slow down and do that in person as well.

"Typical woman!"

Ingressive	Typical man!
Passive	[Silence]
Congressive	That was hurtful.

"It's sweet that you believe in logic."

Ingressive	Too bad you haven't grasped the concept of what logic actually is.
Passive	Well, I guess I'm an optimist.
Congressive	What do you mean by that?

"Women are just vain."

Ingressive	Well, men are just idiots. You could stand to look in the mirror once in a while.
Passive	No, we're not.
Congressive	That was rather belittling.

"You're not very scientific, are you?"

Ingressive	Said the pot to the kettle. So when was your last scientific paper published?

Passive	I'm being perfectly scientific.
Congressive	Which particular part of my argument do you disagree with?

"Congratulations! You earn so little you might as well quit."

Ingressive	Right, and just be a career failure like you.
Passive	But it's better than nothing.
Congressive	Are you trying to be supportive?

"One day you'll learn."

Ingressive	One day you'll retire.
Passive	Maybe.
Congressive	What do you think I'll learn?

"That is not how we do things."

Ingressive	Well, no wonder this place is a disaster.
Passive	Oh, I'm sorry.
Congressive	I'm interested to know what your reasons are.

"Oh yeah? What makes you think you're successful?"

Ingressive	Well, I don't have an insecurity complex like you seem to.
Passive	I think I'm successful in my own way.
Congressive	Do you think there's only one form of success?

"Spoken like a true Chinese."

Ingressive	Spoken like a true racist.
Passive	Huh?
Congressive	What exactly do you mean by that?

Acknowledgments

THIS BOOK HAS been an even bigger and more daunting journey than the previous one, and I was crucially helped by innumerable people along the way, who shared their insight, wisdom, experience, support, and inspiration.

First, I would like to thank Andrew Franklin and all at Profile Books for their deeply congressive ongoing support. Heartfelt thanks are also due to Lara Heimert, TJ Kelleher, and all at Basic Books. I am immensely grateful to have publishers on both sides of the Atlantic who continue to believe in me as I push my boundaries far beyond where I originally saw them. For this book I must again particularly thank my wonderful editor Nick Sheerin, whose vision and empathy combine to help me say what I'm trying to say in a way that will reach further, wider, and deeper.

I owe many thanks to my students at the School of the Art Institute of Chicago. Their intellectual energy and challenging curiosity directly led to this book. I would also like to thank everyone at the School. It is a revelation to work for a supportive, congressive institution.

None of this would happen without the support and inspiration of my parents, my sister, and my little nephews Liam and Jack, who are rather less little than when I thanked them in my previous book.

Thanks are also due to my wonderful friends who have provided my congressive support network for my whole life, and in particular during this phase of my life, in which random insults from around the world are unfortunately quite likely. I could not do all this without that support. Special thanks are due to those whose thoughts and experiences I have specifically referred to in this book: John Baez, Edray Goins, David Kung, Marissa Loving, Ruth Jurgensen, Emily Riehl, Sandmeyer's, and the three hundred or so singers who have thus far sung at the Liederstube. Thanks to Ben Hamill for undertaking some congressive personal training with me.

Thank you to my personal cheerleaders, Amaia Gabantxo and Jason Grunebaum, and my brilliant assistant, Sarah Ponder. As always, I am grateful to Sarah Gabriel for being a beacon to cut through my brain fog.

I would also like to thank all those, too many to name, who have given me an opportunity to speak on this subject at various universities around the world in the last few years while I was developing these ideas. The response from audience members helped me to clarify my thoughts and push them further.

I am grateful for my congressive school education, which celebrated contribution to the community more than "high achievement," although I admit I was a bit bitter about that at the time; I now appreciate the values I learned. I am also grateful to all those who helped my career move forward when I might otherwise not have believed in myself enough: my directors of studies, Paul Glendinning and Jan Saxl; my PhD supervisor, Martin Hyland; and also Paul-André Mélliès, André Hirschowitz, and Peter May.

Last and most, this wouldn't have happened without Gregory Peebles, who has to be known to be believed, and even then is figerally incredible.

Index

Eugenia Cheng is Scientist in Residence at the School of the Art Institute of Chicago and visiting honorary fellow at City University of London. She is the author of *How to Bake Pi, Beyond Infinity,* and *The Art of Logic in an Illogical World.* She lives in Chicago, Illinois.